Enterprise Security

Addison-Wesley Information Technology Series
Capers Jones and David S. Linthicum, Consulting Editors

The information technology (IT) industry is in the public eye now more than ever before because of a number of major issues in which software technology and national policies are closely related. As the use of software expands, there is a continuing need for business and software professionals to stay current with the state of the art in software methodologies and technologies. The goal of the Addison-Wesley Information Technology Series is to cover any and all topics that affect the IT community: These books illustrate and explore how information technology can be aligned with business practices to achieve business goals and support business imperatives. Addison-Wesley has created this innovative series to empower you with the benefits of the industry experts' experience.

For more information point your browser to http://www.awl.com/cseng/series/it/

Sid Adelman, Larissa Terpeluk Moss, *Data Warehouse Project Management.* ISBN: 0-201-61635-1

Wayne Applehans, Alden Globe, and Greg Laugero, *Managing Knowledge: A Practical Web-Based Approach.* ISBN: 0-201-43315-X

Michael H. Brackett, *Data Resource Quality: Turning Bad Habits into Good Practices.* ISBN: 0-201-71306-3

David Leon Clark, *Enterprise Security: A Manager's Defense Guide.* ISBN: 0-201-71972-X

Frank P. Coyle, *Wireless Web: A Manager's Guide.* ISBN: 0-201-72217-8

Frank P. Coyle, *XML, Web Services, and the Data Revolution.* ISBN: 0-201-77641-3

James Craig and Dawn Jutla, *e-Business Readiness: A Customer-Focused Framework.* ISBN: 0-201-71006-4

Gregory C. Dennis and James R. Rubin, *Mission-Critical Java™ Project Management: Business Strategies, Applications, and Development.* ISBN: 0-201-32573-X

Kevin Dick, *XML: A Manager's Guide.* ISBN: 0-201-43335-4

Jill Dyché, *e-Data: Turning Data into Information with Data Warehousing.* ISBN: 0-201-65780-5

Jill Dyché, *The CRM Handbook: A Business Guide to Customer Relationship Management.* ISBN: 0-201-73062-6

Patricia L. Ferdinandi, *A Requirements Pattern: Succeeding in the Internet Economy.* ISBN: 0-201-73826-0

Dr. Nick V. Flor, *Web Business Engineering: Using Offline Activites to Drive Internet Strategies.* ISBN: 0-201-60468-X

David Garmus and David Herron, *Function Point Analysis: Measurement Practices for Successful Software Projects.* ISBN: 0-201-69944-3

John Harney, *Application Service Providers (ASPs): A Manager's Guide.* ISBN: 0-201-72659-9

International Function Point Users Group, *IT Measurement: Practical Advice from the Experts.* ISBN: 0-201-74158-X

Capers Jones, *Software Assessments, Benchmarks, and Best Practices.* ISBN: 0-201-48542-7

Capers Jones, *The Year 2000 Software Problem: Quantifying the Costs and Assessing the Consequences.* ISBN: 0-201-30964-5

Ravi Kalakota and Marcia Robinson, *e-Business 2.0: Roadmap for Success.* ISBN: 0-201-72165-1

Greg Laugero and Alden Globe, *Enterprise Content Services: Connecting Information and Profitability.* ISBN: 0-201-73016-2

David S. Linthicum, *B2B Application Integration: e-Business-Enable Your Enterprise.* ISBN: 0-201-70936-8

Sergio Lozinsky, *Enterprise-Wide Software Solutions: Integration Strategies and Practices.* ISBN: 0-201-30971-8

Joanne Neidorf and Robin Neidorf, *e-Merchant: Retail Strategies for e-Commerce.* ISBN: 0-201-72169-4

Patrick O'Beirne, *Managing the Euro in Information Systems: Strategies for Successful Changeover.* ISBN: 0-201-60482-5

Bud Porter-Roth, *Request for Proposal: A Guide to Effective RFP Development.* ISBN: 0-201-77575-1

Mai-lan Tomsen, *Killer Content: Strategies for Web Content and E-Commerce.* ISBN: 0-201-65786-4

Karl E. Wiegers, *Peer Reviews in Software: A Practical Guide.* ISBN: 0-201-73485-0

Bill Wiley, *Essential System Requirements: A Practical Guide to Event-Driven Methods.* ISBN: 0-201-61606-8

Ralph R. Young, *Effective Requirements Practices.* ISBN: 0-201-70912-0

Bill Zoellick, *CyberRegs: A Business Guide to Web Property, Privacy, and Patents.* ISBN: 0-201-72230-5

Bill Zoellick, *Web Engagement: Connecting to Customers in e-Business.* ISBN: 0-201-65766-X

Enterprise Security

The Manager's Defense Guide

David Leon Clark

Addison-Wesley

Boston • San Francisco • New York • Toronto • Montreal
London • Munich • Paris • Madrid
Capetown • Sydney • Tokyo • Singapore • Mexico City

The publisher offers discounts on this book when ordered in quantity for bulk purchases and special sales. For more information, please contact:

U.S. Corporate and Government Sales
(800) 382-3419
corpsales@pearsontechgroup.com

For sales outside of the U.S., please contact:

International Sales
(317) 581-3793
international@pearsontechgroup.com

Visit Addison-Wesley on the Web: www.awprofessional.com

Library of Congress Cataloging-in-Publication Data

Clark, David Leon, 1960–
Enterprise security : the manager's defense guide / David Leon Clark.
p. cm.
Includes bibliographical references and index.
ISBN 0-201-71972-X
1. Computer security. 2. Electronic commerce—Security measures. 3. Information warfare. I. Title.
QA76.9.A25 C54 2002
005.8—dc21 2002006190

Pearson Education, Inc.
Rights and Contracts Department
75 Arlington Street, Suite 300
Boston, MA 02116
Fax: (617) 848-7047

ISBN 0-201-71972-X
Text printed on recycled paper
1 2 3 4 5 6 7 8 9 10—CRS—0605040302
First printing, August 2002

For Serena, David II, and Stanlyn

Contents

Preface: A Call to Arms

First came Melissa, then Explore.Zip and then the Love Bug. Their names were provocative, fun, and cute. Next came Code Red, Nimda, and, more recently, Reeezak—the triple e's are no typo. Their names, in contrast, are sinister, apocalyptic, and foreboding. So what's in a name? In March 1999, Melissa marked the beginning of the world's reckoning with a new type of Internet virus—a computer worm. A computer worm, a special type of virus, is designed to copy itself from one computer to another by leveraging e-mail, TCP/IP, (Transmission Control Protocol/Internet Protocol), and related applications. Unlike normal computer viruses, which spread many copies of themselves on a single computer, computer worms infect as many machines as possible.

By all accounts, computer worms are nasty critters that have wreaked considerable damage and wasted billions of dollars in computer worker hours. The Love Bug, Code Red, and Nimda cost the Internet community more than $11 billion in productivity and wasted IT staff time for cleanup. The Love Bug alone cost the global Internet community close to $8 billion and eventually infected approximately 45 million e-mail users in May 2000. In July 2001, Code Red cost the Internet community $2.6 billion; in September 2001, Nimda caused $531 million in damage and cleanup.

In January 2002, yet another computer worm, with the somewhat ominous-sounding name Reeezak, unleashed itself on the Internet community. Reeezak, like other worms, appears in e-mail with an innocent-sounding subject: in this case, "Happy New Year." The message of the e-mail—"Hi . . . I can't describe my feelings, but all I can say is Happy New Year ☺ Bye."—comes with an attachment, called Christmas.exe, which when double clicked sends itself to all addresses listed in the user's address book and attempts to delete all the files in the Windows directory and antivirus programs. The worm also disables some keys on the keyboard and propagates itself by using Microsoft's compatible version of IRC (Internet Relay Chat) program. Reeezak, like other worms, affects only users of Microsoft's Outlook or Outlook Express e-mail clients.

If the proliferation of e-mail worms were not insidious enough, the Internet community also experienced the effects of another class of attacks in February 2000, just a few months before the Love Bug. The now infamous and shocking *distributed denial-of-service* attacks on several of the largest and most popular e-business sites—Amazon, Yahoo, eBay, and E-Trade—were not only brazen, making the headlines of many major metropolitan newspapers, but also a wake-up call to the high-flying e-commerce world.

The cumulative effects of successfully orchestrated attacks are taking their toll on the Internet economy. At a minimum, users are frustrated and their confidence shaken. Also, a cloud is raining on the parade marching with fanfare toward e-business horizons. Attacks can be potentially devastating, especially from a financial standpoint. In the case of E-Trade, livelihoods were affected on both sides of the *virtual supply chain,* the new business model that is enabling online businesses to reinvent themselves to capitalize on dynamic e-business marketplaces.

Stock traders who subscribe to the e-commerce service lost the ability to queue up their orders, beginning at 7 AM, so that the trades could be triggered at the start of the opening bell at 9:30 AM. In addition to being livid because legitimate orders were being *denied* by bogus activity flooding the site, the stock traders lost critical financial advantage for certain security tenders. The owners of the breached e-business sites were embarrassed, to say the least. They also inherited a potentially explosive problem that raises the question of security immediately and the viability of e-commerce as a long-term business enterprise. More important, though, customers who lose confidence in their ability to conduct business safely and expediently at these sites will go elsewhere. Lost customers are unmistakably the death knell for Internet enterprises.

The discussion could go on and on with examples, but you get the message. Operating in the Internet economy is risky indeed! So what can be done about it? That is the purpose of this book. *Enterprise Security: The Manager's Defense*

Guide is a comprehensive guide for handling risks and security threats to your internal network as you pursue e-business opportunities. Network security, which factors in *open access* to the enterprise's information assets, is e-business security. Open access allows online transactions to incorporate critical information for customers, suppliers, and partners no matter who they are or where they are.

E-business security is an extension of the security provided by firewalls and virtual private networks (VPNs), integrated with risk management, vulnerability assessment, intrusion detection, content management, and attack prevention. In intranets and extranets and servers in the demilitarized zone (DMZ), firewalls protect the information assets behind their walls. When information is in transit via the Internet, firewalls hand off protection of transactions to VPNs. But when information assets are residing behind the perimeter of firewalls or are not in transit, how do you protect them?

That's the domain of e-security. E-security solutions factor in *scanning technologies* to actively police operating systems, applications, and network devices for vulnerabilities in the infrastructure needed to process, maintain, and store the enterprise's information assets. In other words, e-security solutions identify potential threats, or security events, such as denial-of-service and/or viruses. E-security also provides real-time scanning to detect *in-progress port scans* or intruders looking for an unsecured *window* or door to gain illegal access into your network. After detection, e-security solutions facilitate corrective or preventive action before the attack can be launched, without disruption to the network. E-security also provides a framework for *surviving* an attack in progress.

This book also provides a detailed conceptual review of the most popular detection, assessment, hardening techniques, and real-time security systems that can be integrated to provide life-cycle security solutions. In summary, this book discusses a systematic process of protecting network information assets by eliminating and managing security threats and risks while doing business in the free society of the Internet.

Why This Book

It goes without saying that networks are complex systems and that providing the optimum level of network security has been particularly challenging to the IT community since the first personal computers (PCs) were attached to network cabling decades ago. Today, providing network security could be overwhelming! For a business, the prospect of going online is so compelling primarily because of the pervasiveness of the Internet and the promised payoff of exponential returns. The technologies of the Internet are also a significant drawing card to the business community. The ability to present your information assets in multimedia views is difficult to forgo. Suddenly, it seems that 3-D graphical views, graphics, animation, video and audio functionality, and low-cost communication are the preferred methods of building brand loyalty from consumers or preferred vendor status with customers. These technologies also provide partners and suppliers with a strategic advantage if they are connected directly to critical information assets required for competitiveness and meeting business objectives. The technologies of the Internet also make it easy to collaborate through e-mail messaging and workflow processes and to transfer huge amounts of information cost-effectively.

As easily as these technologies are embraced, however, they are also criticized because of their inherent security problems. TCP/IP is a communications marvel but inherently insecure. When the protocol was a design spec, the creators had no compelling reason to build in basic encryption schemes in the free-spirited operating climate of the computing world when TCP/IP was conceived in 1967. Basic security could have possibly been built in at that time, setting the stage for other systems to be secure when spawned by the Internet decades later. Microsoft's tools and application systems, such as Visual Basic, Outlook, Windows NT, and various office suites, are forever being slammed by disappointed users for the company's apparent decisions to trade off security in order to be the first to market. Even PPTP (Point-to-Point Tunneling Protocol), Microsoft's security protocol for dial-up VPN tunneling, was also fraught with security problems in the beginning.

Even Sun Microsystems's Java, a secure programming language for creating spectacular e-business applications, is not without its problems. And depending on security policy, many enterprises turn applets off in user browsers to prevent malicious code that may be attached to the applets from finding its way into systems when initially downloaded. Therefore, because of the inherent insecurities of Web-enabled technologies, the complexity of the functional aspects of networks, multiple operational layers, and, more important, the skill of hackers, e-security must be *inherently* comprehensive.

Consequently, this book reveals how security must be implemented and administered on multiple levels for effective network security. This book systematically reviews the processes required to secure your system platform, applications, operating environment, processes, and communication links. Effective e-security must also address the tools used to develop your information assets, consisting of applications, programs, data, remote procedures, and object calls that are integrated to present your intellectual capital through the dynamic multimedia world—virtual supply chain—of the global Internet economy.

About This Book

Enterprise Security: The Manager's Defense Guide is a comprehensive description of the effective process of e-security, the human threat, and what to do about it. In intranets and extranets, information assets are defended on the perimeter of the enterprise network by firewalls. Information that traverses the Internet is protected by VPNs and secure socket layers provided by browser-based encryption. But when information is either residing behind the perimeter, perhaps dormant or not in transit, how is it protected? This is where e-security comes in.

The subject matter of this book is presented in four parts. A description of each part follows.

Part I, The Forging of a New Economy, discusses the hypergrowth opportunity the world refers to as e-business. Chapters 1–3 make the case for e-security and why it's a closely connected enabler of e-business, the new economy. Part I also takes you into the world of the hacker, a surprisingly well-organized one. The seriousness of the hacker problem is highlighted, along with a review of how hackers may single-handedly jeopardize the future of e-business as a viable industry. In order for e-business to achieve its expected supergrowth projections over the next several years, an arms race will ensue, with no definite end in sight.

Part II, Protecting Information Assets in an Open Society, discusses the triumphs of firewalls, controlled network access, and VPNs. Chapters 4 and 5 also discuss the glaring shortcomings of these security systems as perimeter and in-transit defenses and point to the need for more effective solutions. In addition, Part II enumerates and discusses the specific security problems that arise if IT mangers rely on perimeter defenses and controlled access alone to protect their enterprise networks. Part II also introduces an overview of complementary methodologies, such as intrusion detection, vulnerability assessment, and content management. When used together with perimeter defenses, these methodologies will provide Web-based enterprise networks with total security, or as much as is practical in the world today. After completing Part II, you should have a greater appreciation of a system of security measures that, when put in place, will effectively thwart hackers, including the malicious ones, or crackers.

Part III, Waging War for Control of Cyberspace, comprises a major portion of the book. In Chapters 6 through 11, you are exposed to how hackers and crackers wage war in cyberspace against hopeful denizens of the new economy. Specific weapons—software tools—are covered, including the distributed denial-of-service (DDoS) tools that brought down E-Trade and effectively disrupted service in Amazon.com and eBay. Part III also presents e-security solutions, which IT managers can deploy for effectively handling the clandestine tactics of the wily hacker. After reading these chapters, you should have a practical knowledge of e-security solutions designed for protecting enterprise networks in the new economy.

Part IV, Active Defense Mechanisms and Risk Management, concludes the book. Chapters 12 and 13 discuss specific processes involved in implementing and using tools and methodologies that provide security for network infrastructures and related applications for e-business. The e-security components of vulnerability and risk management, along with vulnerability assessment and risk assessment and their interrelationships, are covered in full and are carefully positioned as a total solution for deploying security effectively. An extensive set of guidelines is provided such that both the IT and the nontechnical professional can follow. Following these guidelines to implement the total e-security solution will result in fully protecting the enterprise's network against hacker incursions.

Four appendixes provide important details for facilitating the overall e-security process. A glossary and a bibliography are also provided.

Intended Audience

This book is intended for small, medium, and multinational corporations; federal, state, and local governments; and associations and institutions that are intrigued with the potential of the Internet for business opportunity and providing services. Organizations have various reasons to be interested in conducting commerce over the Internet: Competitiveness is one, and improvement of services is another. But the ultimate motivation for this momentum appears to be the monetary rewards associated with effectively harnessing online supply chains for the world's Internet community. In response to such ambitions, organizations are wrestling with the challenge of connecting business partners, customers, suppliers, remote field locations, branch offices, mobile employees, and consumers *directly* online to the enterprise network. Organizations are also wrestling with the risks of allowing open access to information assets. The e-business community requires comprehensive but easy-to-manage security solutions to handle security risks to the enterprise network. If these problems aren't effectively addressed, the outcome could be devastating to the long-term viability of e-commerce.

This book provides a detailed review of e-security, a process of protecting online information assets in the virtual supply chain provided by enterprises over the Internet. E-security incorporates state-of-the-art IT-based security products, methodologies, and procedures for delivering rapid return on investment (ROI), uninterrupted network availability, proactive strategies, barriers to malicious intent, and confidence in the overall integrity of the e-business products and services.

The following types of readers can benefit most from this book.

- *Chief information officers (CIOs)* have decision-making authority and responsibility for *overall* information technology infrastructure and policy for the entire enterprise. Providing secure communications and protecting information assets without disruption to the business process are examples of typical challenges faced by CIOs. In theory, when an organization is involved in an e-business venture, executive IT management already understands the importance of enterprise network security. Chapter 4 should be of particular interest if only firewalls and/or VPNs are in use to protect the network. Chapter 4 discusses the shortcomings of perimeter defenses and points to the need for stronger security measures. Chapter 5 reviews specific security breaches and an overview of e-security's functional framework. Chapter 8 and Chapters 10 and 11 expand on the e-security framework presented in Chapter 5, providing an overview of the functional components of e-security. CIOs should also find Chapters 12 and 13 equally important.
- *Other executives/department managers* may be charged with providing and maintaining the information assets that drive the virtual supply chain of the e-business apparatus. Therefore, Chapters 1–3, which define e-business and e-security and describe the malicious opponents of e-business will be of particular interest. Chapter 1 reiterates the exciting business potential of e-commerce. Chapter 3 discusses the potential barriers that hackers pose to the prosperity of e-business. Chapter 3 is also a chilling reminder that if networks aren't secure, e-business will never reach its full potential. Chapters 12 and 13 are also a must-read for executive managers.
- *MIS/IT managers, Web masters and security professionals,* the main audience for this book, typically have *direct,* or *managing,* responsibility for network security and may also have the unenviable task of translating the business requirements into network security solutions, evaluating the impact of the

new solution on the infrastructure, and implementing and managing the security expansion and process. These topics are the subject of the entire book.

- *System analysts/project managers* too should find the entire book of interest. Chapters 8–11 will be of special interest.

Acknowledgments

- I would like to acknowledge my editor, Mary T. O'Brien, and assistant editor, Alicia Carey, for their patience and professionalism.
- I would like to thank reviewers Anne Thomas Manes, Joshua Simon, Sherry Comes, and Scott C. Kennedy for their critical, in-depth, and thought-provoking comments, suggestions, and insight.
- I would like to thank Stanlyn, my loving wife and soul mate, for her long hours dedicated to editing the book and her gentle encouragement.
- I would like to acknowledge my role models, my three older brothers—James, Christopher, and Michael—for always striving to be their best and part of a greater spiritual whole.
- I would like to acknowledge my younger siblings—Ronald, Dwayne, and Deborah—for their faith in a big brother.
- I would like to thank Doris L. Reynolds, the grandmother of my children and my surrogate mother, for always being there.
- I would like to thank my cousins, Usher A. Moses and Sandranette Moses and family, for helping me to remember my roots, the importance of family, and the inspiration from dreaming together as a family.
- I would like to acknowledge my three best friends—Steven R. Brown, Luther Bethea, and John L. King—for helping me keep it real and to appreciate what's fun in life since our childhood.
- I would like to acknowledge Jackie Jones for being the godmother of my two children, my wife's best friend, and my professional colleague.
- I would like to acknowledge my lifelong friends, Mark and Vera Johnson for being an inspiration, our confidants, and professional colleagues.

PART I

The Forging of a New Economy

It is interesting to speculate on what historians will say about this *revolutionary* era of business. Will they say that we were visionary, opportunistic, and prudent businesspersons pioneering the world to the efficacy of a new business economy? Or will historians look back on this time through jaundiced eyes because the world was driven toward the use of a notoriously insecure global medium in the Internet by short-sighted, greedy, and self-serving entities? Or, were we influenced by individuals who cared little for the long-term viability of the world's international business community, eventually setting the stage for the global apocalypse that the business world succumbed to during a dark era in the future?

Only time will tell. Nevertheless, we are witnesses to a business revolution that rivals the Industrial Revolution of an earlier century. In Part I—Chapters 1 through 3—the phenomenon called e-business is discussed in detail. Chapter 1 takes an in-depth look at the e-business revolution and its tremendous lure to modern-day business entrepreneurs. In Chapter 2, e-security is defined, and its inextricable connection as an e-business enabler is carefully laid out. Chapter 3 explores the clandestine world of the hacker and looks at the political forces mobilizing to thwart the progress of hackers. An arms race is under way for the global Internet economy.

CHAPTER ONE

What Is E-Business?

In this chapter, the e-business phenomenon is defined, or perhaps better stated, its utopian allure qualified. Why are so many businesspersons, entrepreneurs, and investors being seduced, given that the Internet is insecure? More important, what are the implications for security when an enterprise's information machine is connected to the Internet? Further, how does one cross the *digital chasm* from the physical world to a virtual one in order to do e-business? Finally, the significance of virtual supply chains is discussed, along with the effects of critical e-business drivers. The chapter concludes by setting the stage for e-security, the critical success factor in pursuing e-business opportunities.

The E-Business Sweepstakes

Electronic business, or e-business, is the phenomenon that is simultaneously legitimizing the Internet as a mainstream communications medium and revolutionizing a new commercial business reality. The growth potential for creatively conceived and well-managed e-business ventures is unparalleled in the history of industry. *Electronic retail* (e-tail), also known as *business-to-consumer (B2C),* sales were estimated to be more than $12 billion in 1999, with $5.3 billion in the fourth quarter alone, according to official Census Bureau estimates. In a September 1999 study by Prudential Securities, analysts predicted that hypergrowth for e-tail sales would continue into the twenty-first century, beginning with 130 percent growth and leveling off to about 45 percent by 2004. This equates to a compound average growth rate (CAGR) of approximately 69 percent. Prudential Securities research also suggests that annual e-tail sales should reach $157 billion by 2004. Forrester Research predictions are even more optimistic. Forrester estimates that sales resulting from purchases of goods and services through *online stores* will nearly double each year through 2004. In other words, online consumer sales are expected to reach $184 billion in 2004.

Speaking of hypergrowth, *business-to-business* (B2B) e-commerce, whereby businesses sell directly to one another via the Internet, was five times as large as business-to-consumer e-commerce, or $43 billion in March of 1998, according to a report in *Business Week.* Forrester Research predicts that B2B will mushroom to $2.7 trillion by 2004. That's nearly 15 times the size of the consumer e-commerce market projection! In comparison, Gartner Group's predictions are off the chart. The consulting firm expects B2B e-commerce to be almost three times the Forrester prognostication or $7.4 trillion.

Following are some other interesting trends that are fueling the Internet migration.

- Of the 100 million people connected to the Internet, most had never heard of it four years earlier.
- According to an April, 1998, federal government report, "The Emerging Digital Economy," the Internet's *rate of adoption* outpaces all other technologies that preceded it. For example, radio was in existence for 38 years before 50 million people owned one. Similarly, television was around for 13 years before 50 million people were able to watch *American Bandstand.* And, after the first PCs embarked on the mainstream, 16 years were needed to reach that threshold.
- Four years after the Internet became *truly* open to the public—the National Science Foundation released restrictions barring commercial use of the Internet in 1991—50 million individuals were online by 1997. At this rate, especially with 52,000 *Americans* logging onto the Internet for the first time every day, experts believe that 1 billion people will be online worldwide by 2005.
- In spite of the dot-com flameout, companies are still looking to streamline operations by harnessing the Web, according to a June 20, 2001 report in the *Washington Post.*

So at this juncture, the question is not *whether* you should go online but *when* and to *what extent.*

Caesars of E-Business: An Embattled Business Culture

Like the celebrated emperors who ruled the Roman Empire, the new Caesars of e-business are forging business empires through new, *virtual business* channels and as a result are becoming a force at the top of the business world. Loosely defined, an empire is an economic, social, or political domain that is controlled by a *single* entity. Amazon.com, Auto-by-Tel, Beyond.com, Barnes and Noble, CDNow, eBay, and E-Trade are among the new Internet Caesars that appear to be conquering this new *cyber*business world by building an empire in their respective online product or service categories.

Amazon.com became the first *online bookstore* when it hung up its virtual shingle in 1995. In 1996, its first year of operation, it recorded sales of $16 million. A year later, sales had grown nearly tenfold, reaching $148 million. It is estimated that Amazon will realize $2.8 billion in sales in all product categories—books, CDs, movies, and so on—in 2003!

Amazon's literal overnight success became too compelling to pass up. Barnes and Noble, a bricks-and-mortar establishment, set up its own online shop to compete in the seemingly fast-growing book market in 1997. Online book sales are expected to reach $3 billion by 2003.

Most industry analysts are ready to concede the online book empire to Amazon and Barnes and Noble. Through Amazon alone, its 11 million customers can select from more than 10 million titles, consisting of 1.5 million in-print books in the United States and 9 million hard-to-find and out-of-print books.

On other online product retail fronts, Beyond.com is building its business empire in the *online software sales* category, with more than 48,000 software application product titles. Similarly, CDNow offers more than 325,000 CD titles to its online customers, and eBay has locked up the *online auction* front for trading personal items of wealth. Amazon.com and eBay are well on their way to building business empires, perhaps reaching that coveted milestone of category killers for book sales and auction trading, respectively (see Table 1–1).

Feeling the effects of Barnes and Noble's actions, Amazon responded with incisive moves into other areas. In June 1998, Amazon.com opened its music store, going head to head with CDNow. This move was followed by a rollout of virtual toy and video stores, positioning Amazon.com for direct competition with eToys and Reel.com, respectively. Amazon didn't stop here. It also set up shop in the online greeting cards, consumer electronics, and auction areas. Within 90 days of launching its music store, Amazon became the premier online

Table 1-1 Competitors in the Online Market Segments (Product Categories)

Product Category (Market)	Potential Category Killers		
	Original E-Tailer	E-Tailer Crossover	Traditional Retailer
Books	Amazon.com	Buy.com	Barnes and Noble (Bn.com)
Music (CDs, etc.)	CDNow	Amazon.com	Tower Records
Videos	Reel.com	Amazon.com	Blockbuster Videos
PC hardware	Buy.com	Egghead	CompUSA, Dell, Gateway, Compaq
Toys	eToys	Amazon.com	Toys-R-Us, Wal-Mart, KayBee
Software	Beyond.com	Amazon.com Bn.com	CompUSA, Egghead
Autos	Autobytel.com, Cars.com, Autoweb.com	N/A	Harley-Davidson
Consumer electronics	800.com	Amazon.com	Best Buy, Circuit City

music retailer; within 6 weeks of launch, the premier online video retailer. Not to be outgunned, CDNow reciprocated by opening online movie and book businesses. Other online retailers began following this strategy.

No sooner than the *online* giants begin moving in on one another's turf, the *traditional* retailers begin to exert their *physical* muscle in the virtual world of compelling shopping malls and online stores. Blockbuster set up a Web site to sell movies. Toys-R-Us raised no eyebrows when it decided to go online to challenge eToys in the online toy category. Tower Records moved into CDNow's and Amazon's territory to challenge in the music arenas. The incursions of the online retailers and the invasions of the traditional retailers make for a crowded virtual marketplace, indeed.

The Lure Of Overnight Successes

While the mega-e-tailers were jostling for control of their respective online empires, roughly 30,000 e-tailers sprang up like Christmas lights to ply their wares through the Web. The overnight success of Amazon, Barnes and Noble, Dell Computers, Auto-By-Tel, and other Internet retailers was an intoxicating lure to opportunistic Internet entrepreneurs looking to capture that magic formula. Unfortunately, dot-coms failed by the thousands. In fact, in the fourth quarter of 2000, industry analysts predicted that more than 80 percent of e-tailers, or 25,000 companies, would not succeed in the cutthroat online retail business. Those that were absorbed by bigger concerns were fortunate, to say the least. However, the debacle of the dot-com businesses and other adverse market forces impacted high-tech stocks in general, causing stocks in other high-tech areas, such as Microsoft and Cisco, to sustain a decline in market value.

The five-year period ending December 2001 saw Internet giants completing their initial public offerings (IPO) and entrepreneurs, management, venture capitalists, and other investors who were holding stock options become overnight millionaires, even billionaires! Amazon completed its IPO on May 15, 1997, after opening its virtual doors in July 1995. The stock price reached $113 a share in December 1999! A year later, the stock was trading at approximately $20 a share; by December 2001, $10 a share! This is truly phenomenal, given the fact that Amazon has been in operation for only six years. Even more amazing, as the dot-com shakeout continues, forecasters are expecting solid growth in all online product categories. The failings of the dot-coms and the debacle of high-tech stocks were inevitable, if not expected. Some industry analysts point out that the recent adversity is a natural correction of a marketplace, which is returning to equilibrium. The overvalued capitalization, inflated stock prices, and exponential returns from the IPO have simply run their course.

Oddly enough, investors quickly understood that to play in the online retail game, an infusion of capital would be needed to develop online business models successfully. In general, virtual supply chains represent online infrastructure and related processes that harness the attributes of the Internet for the purpose of delivering goods and services, emulating physical supply chain infrastructure and processes of *traditional* retail with software application processes and network infrastructures for *online* retail. The challenge for online retailers is to craft an automated business system that will garner success online. Investors, betting that several years of heavy capitalization will ultimately achieve acceptable returns in the foreseeable future, are therefore willing to live with substantially undervalued stock prices in the near term for riches in the future. Besides, investors who held onto their shares since the IPO have made and lost a ton of money.

Without doubt, the mystique and the attraction of the Internet as a viable business channel have been glorified and substantiated by the innovative pioneering of the super-e-tailers, the Caesars of the Internet economy. But as mentioned, business-to-business e-commerce is expected to be 10 to 15 times larger than the retail online business. Moreover, companies collaborate over the Internet for purposes other than direct selling, such as to exchange information with employees or strategic business partners. Thus, companies interacting online to provide products and services directly or to gain strategic and/or competitive advantage realize the fullest, perhaps the most practical, intent of the Internet. How this will be achieved from company to company will vary significantly.

Crossing the Digital Chasm

No matter what e-business model you choose—B2C, B2B, an intranet for internal use, or an extranet for strategic external entities, such as business partners—you must fashion the requisite computer application(s) in order to pursue e-business opportunities successfully. To qualify as an e-business application, it must allow access to the intellectual capital, or information assets, of the enterprise while operating safely on the Internet. In general, e-business application development depends on four critical factors: where information assets reside, how they are processed, who manages the application, who is beneficiary; in short, the database, applications, IT/operating staff and the end user (see Figure 1–1). Critical e-business drivers include streamlining physical operating processes, reducing operating costs, delivering just-in-time information, and increasing services to customers (see Figure 1–2).

No matter how you slice it, the development of e-business applications is not a walk in the park. Internet-enabling technologies facilitate the achievement of this end and even make it fashionable. However, determining which of the vast amounts of information capital you deploy for a given e-business application may be a straightforward process or as complicated as enterprise application integration (EAI). EAI is a process that identifies and integrates enterprise computer applications or databases, typically in dissimilar formats, into a derivative, or new, computer application using middleware models and related technologies such that the resulting application is accessible through a graphical user interface (GUI).

The critical first step in e-business application development is deciding *what* business activity would be more effective as an e-business application. In

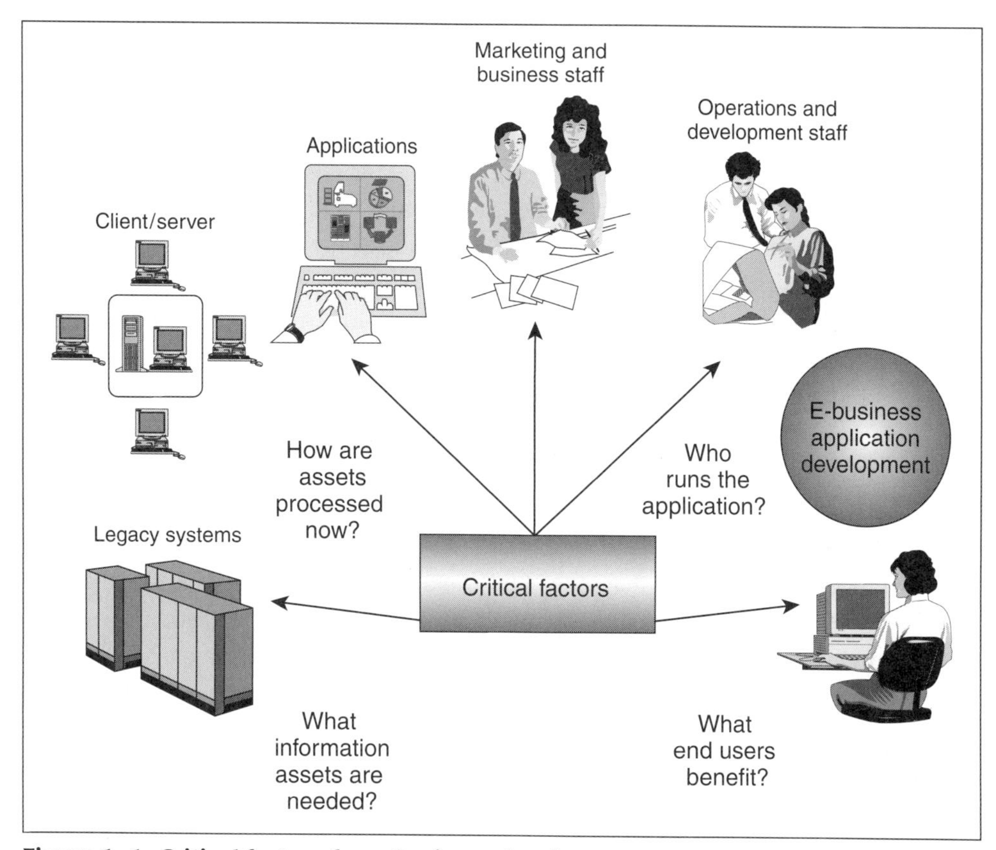

Figure 1–1 Critical factors for e-business development

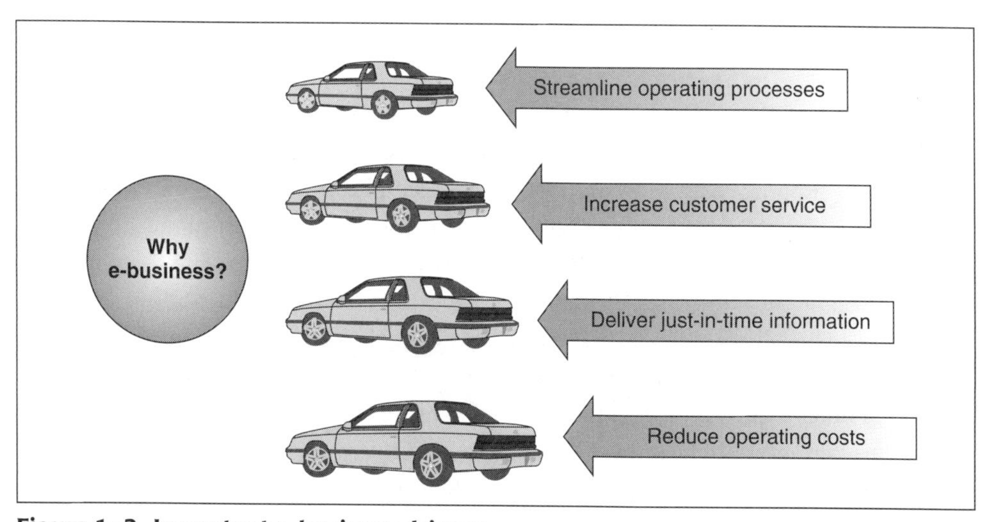

Figure 1–2 Important e-business drivers

its simplest form, e-business involves incorporating the Internet or its technologies to support a *basic* business process. For example, your order entry system, connected directly to the inventory database, is typically accessed from the field by sales reps calling their *product availability* inquiries in to an order entry administrator. The sales reps call in through a static GUI program or by e-mail to an order entry clerk, who processes each inquiry by order of receipt. The process works but may bog down during peak periods of the day or when the staff is short-handed. Besides, the main function of the order entry staff is to process *actual* orders. Providing product availability information to the field is a related responsibility that is often superceded by higher priorities. Processing last-minute requests in preparation for a meeting is too often out of the question. To complicate matters, you also have independent dealers and affiliates requiring product availability status reports as well as inquiries on an ongoing basis.

After deciding that the product availability inquiry activity is suitable for an e-business application, the next step is identifying the information asset(s) the process generates. The mapping of information assets with the processes that support them is a critical requirement in e-business application development. In this example, the information asset created by the process is "product availability" (see Figure 1–3). After receiving the inquiries, the order entry staff queries the inventory database to check the status of products from key suppliers. When the availability of a particular product is ascertained, the information is conveyed back to the end user via e-mail or fax. The product availability information allows sales representatives to respond to clients effectively. Finally, you recognize that the order entry staff performs a clearinghouse function, or *a physical (manual) process,* which ensures that inquiries and responses are cleared out of the queue.

To be most effective, the e-business application would have to provide up-to-the-minute information to field personnel, consultants, and partners and also eliminate or streamline the product status and clearinghouse function, reducing sales support costs. Moreover, the resulting application would reduce communication costs, given that the Internet replaces traditional communications links, and end users' learning curve would be less, as the system would be accessed through the familiar environment provided by Web browsers.

This all sounds good. However, it's easier said than done. In order for the e-business application to provide the functionality of the previous system, the product inquiry and *physical* clearinghouse process is enhanced by a *digital process,* or computer application. The database—in this case, the inventory database—must also be available and interconnected to the virtual process, or application.

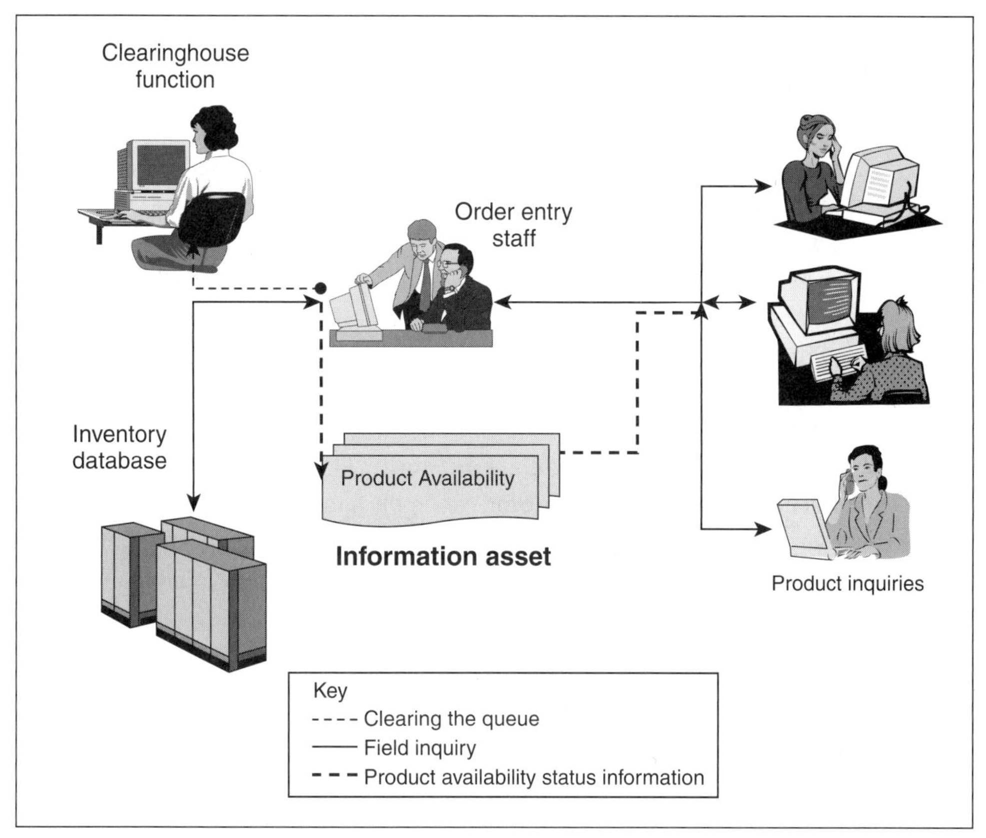

Figure 1–3 Product inquiry fulfillment process

Instead of *field* personnel interacting with a character-based, static GUI or other generic front end to generate the inquiry request, they would access a front end that is capable of running in their browser, a personal digital assistant (PDA), or wireless hand device. The front end—Web server—must be able to perform the function provided by the order entry staff. That is, it must be able to access the inventory database, gather the information required by the inquiry, format the response, and feed it back via the Internet to the appropriate place (field) in the user's browser, which is running the application on a laptop, home office computer, PDA, and so on. The application also does some housekeeping chores by clearing the inquiries from the front end and the remote database calls from the back end, or inventory database.

Most likely, the front-end Web application, or what the users see and interact with in the browser, is developed with Internet-enabled technologies, such as Java or HTML application tools. The back end could be, for instance, a legacy UNIX database that has been a mission-critical application for some time. To accomplish the *interconnectivity* between the front-end browser application and the back-end UNIX database, yet another application system, typically referred to as *middleware,* must be used to provide the interconnections, or compatibility, between the dissimilar front- and back-end applications. Examples of middleware are systems developed with J2EE (Java 2 Platform Enterprise Edition). Developed by Sun Microsystems, J2EE is more popular in Web application development than CORBA (common object request broker architecture), introduced by the Object Management Group in 1991, or DCOM (distributed component object model), which is Microsoft's bet for an object standard. However, the other standards are growing in use for Web application development. With middleware in place, the e-business application provides the same functionality of the previous system. However, the virtual process replaces the traditional product inquiry and physical clearinghouse process and provides greater operating advantages and overall benefits to the enterprise (see Figure 1–4).

You can see that for even the simplistic example shown in Figure 1–4, crossing over from a traditional process to a virtual process to achieve e-business goals could pose a potentially complicated challenge, like crossing a chasm on a tightrope. Crossing this *digital* chasm to pursue e-business opportunities therefore requires a complete knowledge of the enterprise's information assets, or more appropriately, where the necessary information assets reside to support a given e-business application. This crossover also assumes the incorporation of a dynamic, browser-compatible front end and the identification or development of the static back end: the database. Perhaps the most critical aspect of the entire process is deploying the middleware that ties the whole e-business application together. This is the lifeblood of e-business.

The Sobering Reality

As e-business legitimizes the Internet as a mainstream business facility, many individuals have begun to see the Internet more as a basic utility, not a mere convenience. Livelihoods in every field of endeavor are increasingly going online. And when livelihoods are involved, a sense of security is usually an accompanying factor. As previously suggested, the World Wide Web consists of highly complicated yet fallible technology. In dealing with computer networks, a modicum

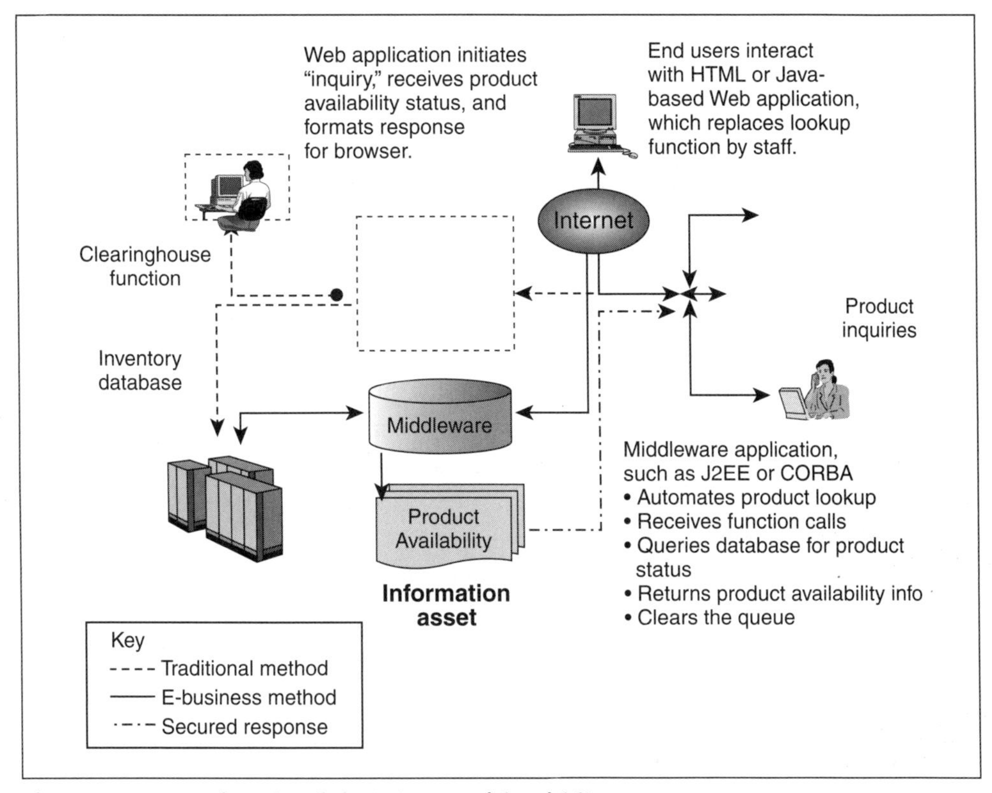

Figure 1–4 Crossing the digital chasm with middleware

of inconvenience is acceptable. Sites get overwhelmed and clogged with traffic, Web servers break down, HTTP and Java applications crash, and huge file transfers affect overall network performance. In general, such events occur without any interference from external hackers and crackers or internal saboteurs. Besides, no one is naïve enough to expect uninterrupted service just because essential applications are moved online. Those occasional hiccups in network service are not usually a threat to our sense of security.

However, as more and more businesses and entrepreneurs make that all-important leap-of-faith in search of increased revenues, operational efficiencies, cost savings, and/or strategic advantages, rest assured that hackers, crackers, and saboteurs will attain more powers of destruction. Fortunately, such powers are not omnipotent enough to stop the momentum of the Internet migration. But they are powerful enough to shake that sense of security we need to pursue our livelihoods. Internet denizens should condition themselves to expect visits from these human-driven menaces.

Real-World Examples

If you want to know what it's like to weather a horrendous storm, just ask E-Trade. E-Trade, the nation's second-largest online broker, pioneered the radical shift from traditional brokers to trading stock online. About 7 AM in early February 2000, E-Trade came under a massive denial-of-service attack. It was no coincidence that the attack began precisely when E-Trade's customers, online brokers, and day traders begin flooding the site with legitimate orders for stock purchases. Much to everyone's chagrin, the site was being flooded with bogus queries, which succeeded in choking the system and at the same time *denying* legitimate subscribers entry to the site. The relentless onslaught of bogus activity continued well after 10:00 AM, successfully locking out business activity during the stock market's busiest time of the day.

In the aftermath of the attack, about 400,000 traders, about 20 percent of E-Trade's client base, were either unable to make trades or lost money owing to the length of time required to complete them. As a stopgap, E-Trade routed some investors to live brokers. Consequently, E-Trade lost millions of dollars when it was forced to compensate traders for losses from trades taking longer than usual and to pay the fees from the live brokerage houses.

A few days before the attack on E-Trade, Yahoo and Amazon.com were also temporarily crippled by denial-of-service attacks. As the now infamous attacks were under way, the Internet economy was stunned, and a sense of helplessness permeated the virtual community.

The attacks bring into focus the shortcomings of the Internet. Although industry observers feel that the attacks will not stunt the exponential growth of the Internet, they highlight the vulnerabilities of the millions of computer networks that delicately link the new economy. Some observers try to equate those attacks with the equivalent of spraying graffiti on New York's subways. Others maintain that real ingenuity and solid citizenship will ultimately win the battle for the Internet's safety and integrity. Such ingenuity could lead to dispensing a host of innovative controls to patrol the freeways of the Internet. In the meantime, business will be conducted but not quite as usual. This era is marking the end of Internet innocence. If you are involved either in e-business or in planning for it, you should condition your expectations for hacker exploits, much like we are conditioned for junk mail, rush-hour traffic, or telemarketers. In the meantime, a gold rush is under way. Although every stake for e-business will not find gold, the *virtual forty-niners* will not be deterred in their mad rush for e-business.

E-Business: The Shaping and Dynamics of a New Economy

E-business is a revolution: a business existence based on new models and digital processes, fueled by hypergrowth and new ideals. It is also pursuit of new revenue streams, cost efficiencies, and strategic and competitive advantages spawned by virtual business channels. Cutting-edge Internet technologies and new vistas of emerging technologies enable e-business. E-business is a forging of a new economy of just-in-time business models, whereby physical processes are being supplanted by virtual operating dynamics. Yes, e-business is all this. But still, what *is* e-business? In other words, what is the intrinsic nature of e-business?

The E-Business Supply Chain

Typically, e-business is described and discussed with more emotion than other business areas, and rightfully so. After all, we are witnesses to an exciting revolution. To gain true insight and a conceptual understanding of e-business, it needs to be defined from both the B2C and the B2B perspectives. This section also introduces Internet, or digital, supply chains and reveals their underlying significance to both the B2C and B2B e-business channels.

The Business-to-Consumer Phenomenon

When consumers purchase goods and certain classes of services directly from the Internet, online retailers are servicing them. In other words, online retailers, or e-tailers, have initiated a consumer-oriented supply, or value, chain for the benefit of Internet consumers. This form of Internet-based activity is known as business-to-consumer (B2C) electronic commerce. In this discussion, supply chain is used interchangeably with value chain. However, supply chain, in the traditional sense, refers to the supply and distribution of raw materials, capital goods, and so on, that are *purchased* by a given enterprise to use in manufacturing or developing the products and services for customers or in regular business operations. In B2C distribution modes, supply, or value, chain refers to the system, or infrastructure, that delivers goods or services directly to consumers through Internet-based channels. But what exactly is B2C e-commerce? But more important, why has it grown into a multibillion dollar industry?

To begin in the abstract, B2C e-business is a rich, complex supply chain that bears no *direct* analogy to the physical world. In fact, *no supply chain* in the physical world compares to B2C value chains such that an apples-to-apples comparison can be made. Thus, B2C e-channels are unique because they are providing supply chains that *streamline and enhance processes* of the physical world (see Figure 1–5). Internet-driven supply chains depend heavily on the coordination of information flows, automated financial flows, and integrated information processes rather than on the physical processes that traditionally move goods and services from producer to consumer.

Three classes of B2C value chains make possible the following e-business realities:

1. Delivery of the *universe,* or an unlimited number—potentially millions—of goods and services within established markets, by operating under a single brand identity or as a superefficient intermediary
2. Creation of *new market channels* by leveraging the Internet
3. Elimination of middlemen while streamlining traditional business processes

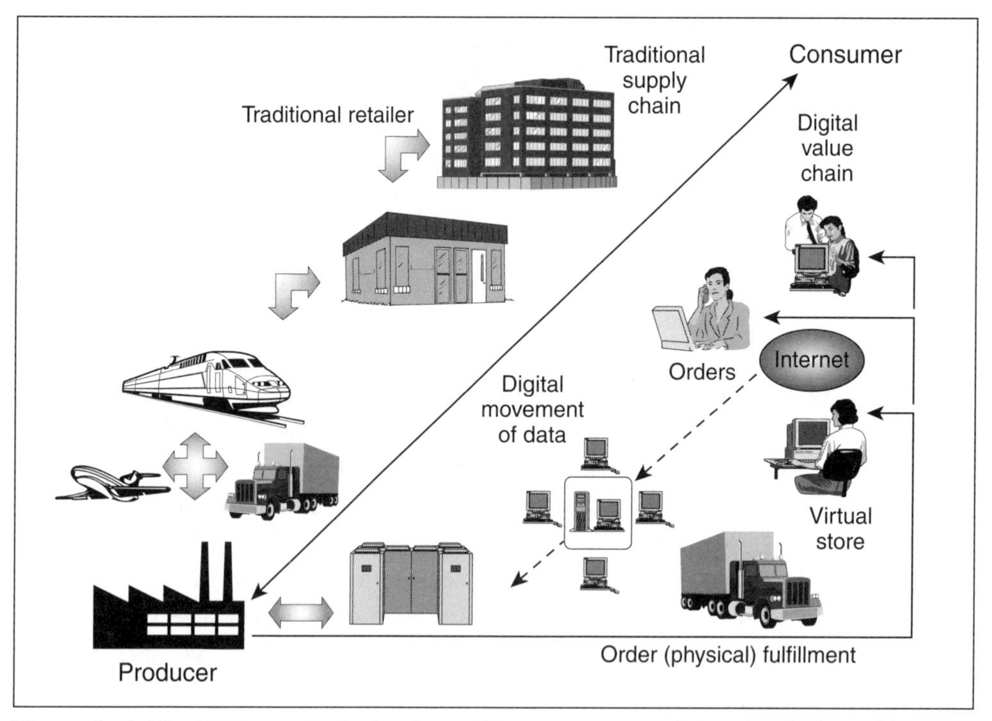

Figure 1–5 The B2C supply chain streamlines processes of the physical world

Amazon.com and CDNow are excellent examples of the B2C class indicated in class 1. Amazon has succeeded by producing an efficient consumer product delivery system. The value in this e-business channel is the uniting of many back-street dealers under the banner of one popular brand name. CDNow is also attempting to implement a similar strategy. Furthermore, no one bookstore or music store in the physical world offers 10 million titles like Amazon.com does or 325,000 CDs like CDNow does. Traditional book or CD retailers in established markets could never offer this vast array of merchandise, because of shelf space and inventory constraints. For example, the typical superbookstore or music CD store stocks only 150,000 or 60,000 titles, respectively.

An example of B2C class 2 is eBay, which created a *new* market channel in establishing an online auction facility. Through this e-business channel, buyers and sellers—everyday consumers—can interact to sell personal items in a venue that did not exist previously.

Dell.com is an example of the third B2C e-business class. Dell.com is successful because it incorporates the principle of *disintermediation,* or the ability to eliminate intermediaries from the value chain. In other words, disintermediation involves disengaging middlemen, who usually command a share of the value chain. Research has shown that intermediaries add a large percentage to the final price of products. Percentages range from 8 percent for travel agents to more than 70 percent for a typical apparel retailer. Dell is a business case example of effective deployment of disintermediation because its direct consumer model delivers *custom-built* computer systems at reasonable prices by leveraging Internet channels. In the future, other online supply chains will successfully remove middlemen, resulting in even lower prices for other classes of goods and services.

Perhaps the common denominator of all three categories is the potential to streamline physical operating processes in the supply chain. This is another important reason that B2C growth through the Internet is so compelling. Physical retailers are capital intensive. When the shelves are fully stocked, adding new products may prove to be too challenging, possibly requiring either displacing more established products or engaging in a costly physical expansion. On the other hand, the incremental cost of adding new products for an online retailer is minimal, especially because the product manufacturer or distributor may carry the inventory. Also, online retailers do not have to incur the cost of operating a showroom floor.

Similarly, the processes of other consumer-oriented services, such as travel agencies, can be streamlined by automation and the overall service provided through the Internet. Such trends serve to pass on the cost efficiencies to consumers, who in turn pay lower prices. Expect to see more service-oriented interests, such as financial institutions, provide more services online in the future as they continue to identify physical business processes that can be enhanced by a move to the digital world of the Internet.

In summary, the Internet supply chains created to support B2C e-business initiatives have no direct analogy in the traditional, or physical, world of commerce. True, the two channels have similarities. The goods and services offered in physical bricks-and-mortar retailers become sexy multimedia presentations and *transaction data.* E-tailers and consumers connect via Web portals instead of driving to malls or to various business concerns. Inventory becomes online transaction data that flows from the consumer's *shopping cart* of the online store—Web site—to fulfillment houses or directly to the producers themselves.

To recap, B2C value chains create the following three types of e-business realities:

1. In established markets, creation of *digital supply chains* that *eliminate middlemen* and enable the availability of a unique service, such as Dell's direct delivery of custom-built PCs.
2. Creation of a *new market channel* that *did not exist* in the physical universe, such as eBay's creation of the online auction facility for the convenience of everyday consumers.
3. *Uniting* of back-end, used or rare-product dealers under the banner of a popular name brand. In effect, this creates a consortium of businesses under a *single branded identity,* or under a new, superefficient intermediary, that *did not exist* in the physical world.

If you can create a B2C value chain that eliminates middlemen, establishes a new market channel for a novel idea, or creates a superintermediary providing an unlimited number of products while streamlining physical processes in all cases with Internet applications, you just might become the next Dell, eBay, or Amazon: a *dot-com.*

The Business-to-Business Phenomenon

B2B is the poster child for e-business. As exciting and awesome as B2C and other e-business opportunities are, they pale next to B2B projections. (See the section The E-Business Sweepstakes earlier in this chapter.) Although prognostications vary across the board, all estimates are in the trillions of dollars. One forecast has

B2B commerce growing from $150 billion in 1999 to $7.4 trillion by 2004! Presently, the median transaction for B2B sites is three to four times the size of the median transaction for B2C sites, or $800 versus $244. Important drivers of this projected growth include, but are not limited to, competitive advantage, reduction of costs, increased profits, and customer satisfaction. If you are able to build an effective B2B channel, the payoff could be significant, resulting in improved economies of scale and productivity, reduction in overhead, improved information flows and processing, and increased operating efficiencies, to name a few.

In light of projected growth, we should expect an exciting, evolutionary time in the development of dynamic B2B Internet channels, ultimately leading to robust extranets that consist of dynamic e-trading communities. At the heart of e-business transformations will be the Internet-enabled supply chain. In fact, you might say that these new digitally oriented supply chains are at the epicenter of e-business migration. Through the benefits of properly implemented B2B supply chains, enterprises can

- *Reduce costs of goods and services and potentially lower customer prices.* By connecting information systems directly with suppliers and distributors, organizations can realize more efficient processes, resulting in reduced unit costs of products or services and, perhaps, lower prices to customers while effectively achieving economies of scale.
- *Reduce overhead.* B2B channels can eliminate extraneous or redundant business functions and related infrastructures, resulting in the reduction of overhead costs.
- *Increase productivity.* By eliminating operational waste and the automation of inefficient business practices, organizations can realize productivity gains.
- *Enhance product and service offerings.* With economies of scale, reduction of overhead, operating efficiencies, and lower operating costs, such gains may be passed on to the customer through lower prices or as enhanced or additional features of products or services.
- *Customer satisfaction.* A strategic benefit of the successful implementation of dynamic B2B business models is improved customer perception of the transaction.

This metamorphosis will not occur unless companies undergo radical changes. Enterprises will begin with critical self-examination and comprehensive process analysis to determine what internal operating functions, underlying infrastructures, and critical practices are necessary to transform into a B2B channel

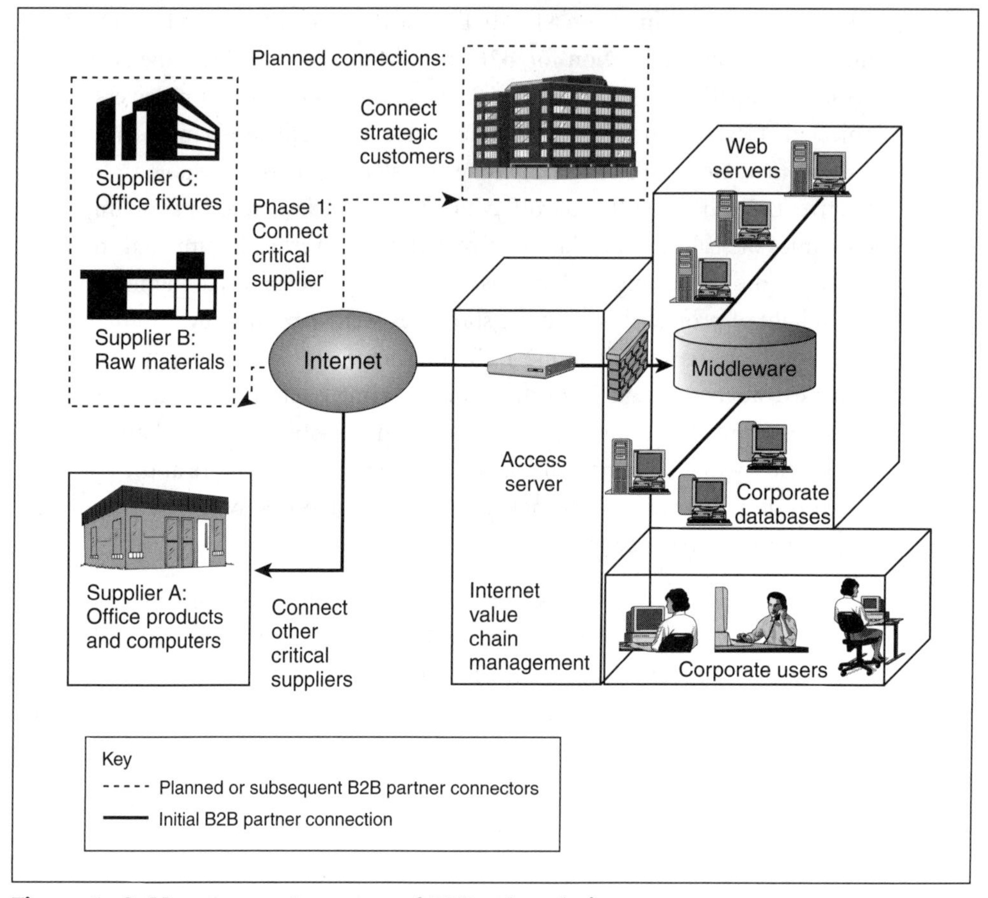

Figure 1–6 Management process of B2B value chain

that is capable of leveraging the Internet. This in turn will lead to the reengineering of processes, elimination of operational inefficiencies, and, ultimately, increased productivity. If companies are successful, they will reinvigorate their value chains, incorporate technology-driven processes that become the foundation for B2B, and increase transactions with customers (see Figure 1–6).

Related E-Business Trends

Pursuing business-to-consumer markets, especially if the goal is to become a coveted *category killer* in one or more product areas, would be too much of an uphill battle for many enterprises. Category killers are the dominant player or players in an online product category, such as Amazon.com and Barnes and

Noble for online book sales. The business-to-business arena may offer greater opportunity for capitalizing on e-business applications. However, the potential IT investments required, along with the general complexity of EAI to achieve seamless B2B information systems could prove beyond the means of many enterprises. But make no mistake about it: E-business is here to stay. The reason is that traditional business processes are being reinvented into new digital processes within the e-business value chain and enabled by Internet technologies. A properly constructed B2B supply chain can reduce costs, increase productivity, enhance product and service offerings, and increase customer satisfaction.

Summary

B2C and B2B business models provide progressive enterprises with the means to reinvent their organizations, streamline business processes, automate traditional business processes, and quickly adapt to new situations, opportunities, and market demands as they unfold. In the rush to embrace e-business, organizations understand that new opportunities must be pursued at the speed of information, which in turn is being enabled by the rich but inherently insecure technologies of the Internet. To be successful, therefore, you must show that Internet supply chains are dependable. If your public Web site is designed to support the overall B2B effort also, acceptable security measures should be implemented there as well.

Another critical success factor is that B2B channels must be more available than traditional competitors, which operate in the physical world. But to meet or to exceed e-business objectives, your suppliers, distributors, and employees must perceive that the B2B online supply chain is secure enough to support the required level of transactions on an ongoing basis. These are the keys to success in the new business reality of B2B value chains.

CHAPTER TWO

What Is E-Security?

To foster buy-in and support of e-business applications from your customers, suppliers, partners, and employees, they must perceive that a reasonable level of security is prevalent, suggesting that the security goals, which include integrity, privacy, and confidentiality, are being fulfilled within the application. *Integrity* says that unauthorized modifications to your information are thwarted as the information resides in the system and travels from point to point. *Privacy* suggests that information is always confidential to the parties directly involved in the communication. *Confidentiality* means that information is protected and not shared with unauthorized parties.

Implementing security measures to meet the requirements of e-business applications presents a distinct set of challenges. Installing appropriate security measures is generally a daunting process, even when networks are closed. So what are the implications for open-access networks, or e-business channels? Enter e-security. Chapter 2 introduces the topic and explains why it is an important enabler of e-business. Moreover, two important concepts are covered in detail: the principles of e-security and the dilemma of open access versus asset protection.

E-Security at Your Service

Achieving confidentiality, authentication, and data integrity was a milestone, a momentous development for networks when the objective was to limit or to prevent outsider access to enterprise information assets and resources. E-business, on the other hand, mandates open access to intellectual capital and information assets by outsiders. Moreover, e-business depends on the seamless interaction of complex multiple computer environments, communication protocols, and system infrastructures among customers, partners, suppliers, branch offices, and remote or mobile employers. Therefore, security that accommodates open access to information assets among heterogeneous networks is e-security. E-security must also be scalable and capable of being administered holistically to the resulting Intranet, extranet, or Web site on an ongoing basis. Consequently, as new needs arise, the security infrastructure must adapt and grow to meet the changing circumstances.

Demands on Traditional IT Security: A Changing of the Guard

To sustain e-business initiatives, e-security must be inherently comprehensive and able to deliver security on multiple operating levels in corporate Web sites, intranets, and extranets (see Figure 2–1). By themselves, firewalls and VPNs (virtual private networks) do not qualify as e-security but rather are considered point solutions. To be fair, however, VPNs and firewalls are dynamic security measures and are good at what they do: protect all points on the perimeter (firewalls) and all information transmissions between security points (VPNs).

However, the functionality of point solutions is based on security models derived from the physical world. Security models in the physical world generally

- Control who has access to a location and where they can go once inside, by means of firewalls and file and directory permissions
- Ensure that information is properly protected when it travels from one trusted location to another, using in-transit point-to-point protection, such as VPNs, SSL (Secure Socket Layer), and encryption
- Make certain that no one is using someone else's keys and that the individual is in fact who he or she is supposed to be, using such authentication measures as smart cards or tokens, strong passwords, and biometrics

Point solutions foster trusted networks, and their effectiveness ultimately depends on their remaining private or virtually private. In other words, as long as the network's devices, hosts, servers, users, and domains are authentic, protected, have data integrity, and *remain closed* to outsiders, the network is secure. Virtually private networks, which protect information transfer over untrusted networks, such as the Internet, protect data with tunneling protocols, encryption, and data

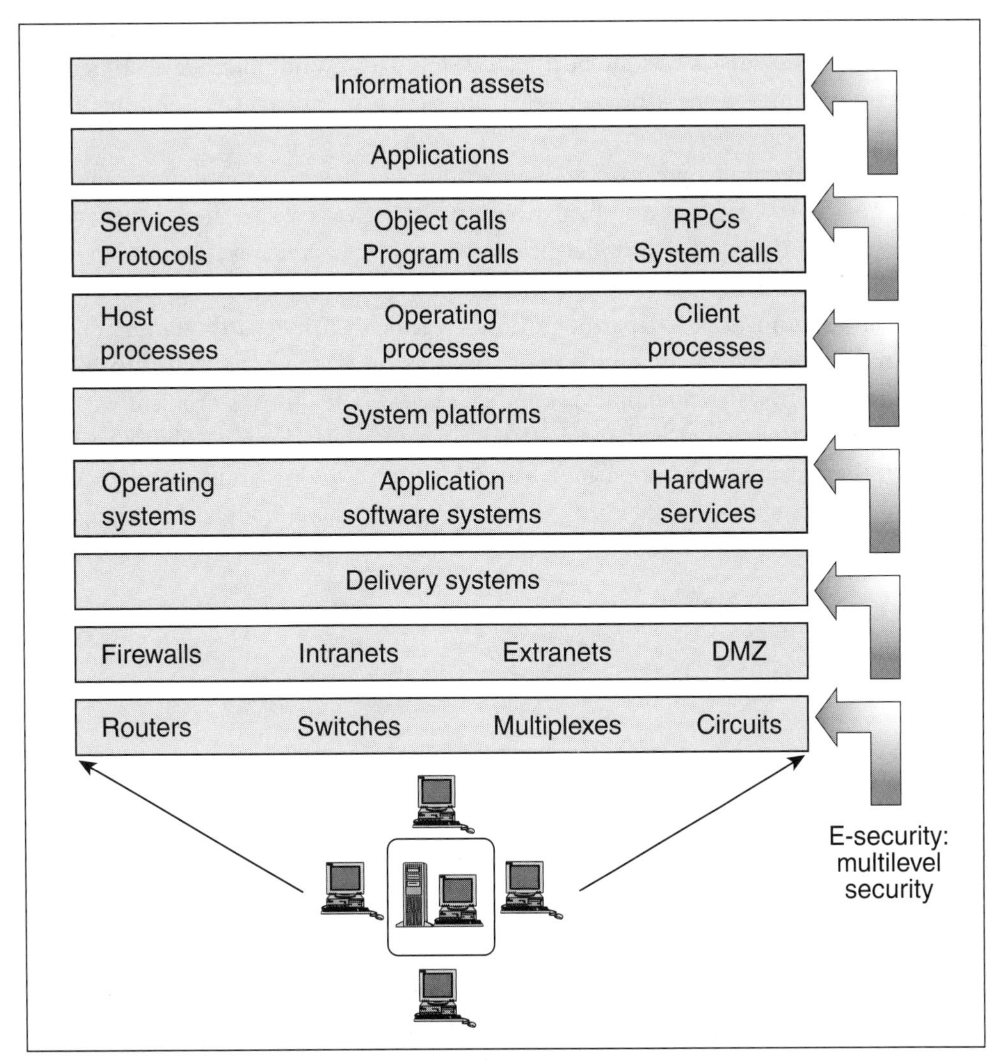

Figure 2–1 E-Security: security for multiple operating levels

integrity algorithms called digital signatures. VPNs also incorporate the use of strong authentication, that is, authentication that requires a password and a personal identification number (PIN). Together with encryption and digital signatures, information is tunneled safely through the Internet.

Achieving competitive advantage translates to keeping costs down yet creating the information needed to make just-in-time decisions. Connecting business partners, suppliers, customers, and vendors directly online to the enterprise's information assets was the next logical step. This evolutionary development, or just-in-time business model, began impacting point security solutions in an adverse way.

Theoretically, point solutions are sufficient to protect e-business environments, provided that a sufficient number of firewalls and related security measures are deployed in a sufficient number of places, creating sub- or concentric perimeters within an overall perimeter (see Figure 2–2). An analogy in the physical world might be a department store's placing security guards in every department, security tags on all the goods, sensors on all the doors, and surveillance cameras to watch all employees and customers. The obvious problem with this type of deployment is that point security solutions lack the flexibility required to be practical in an e-business environment. More important, reliance solely on point solutions would prove to be too expensive and not scalable enough for e-business applications.

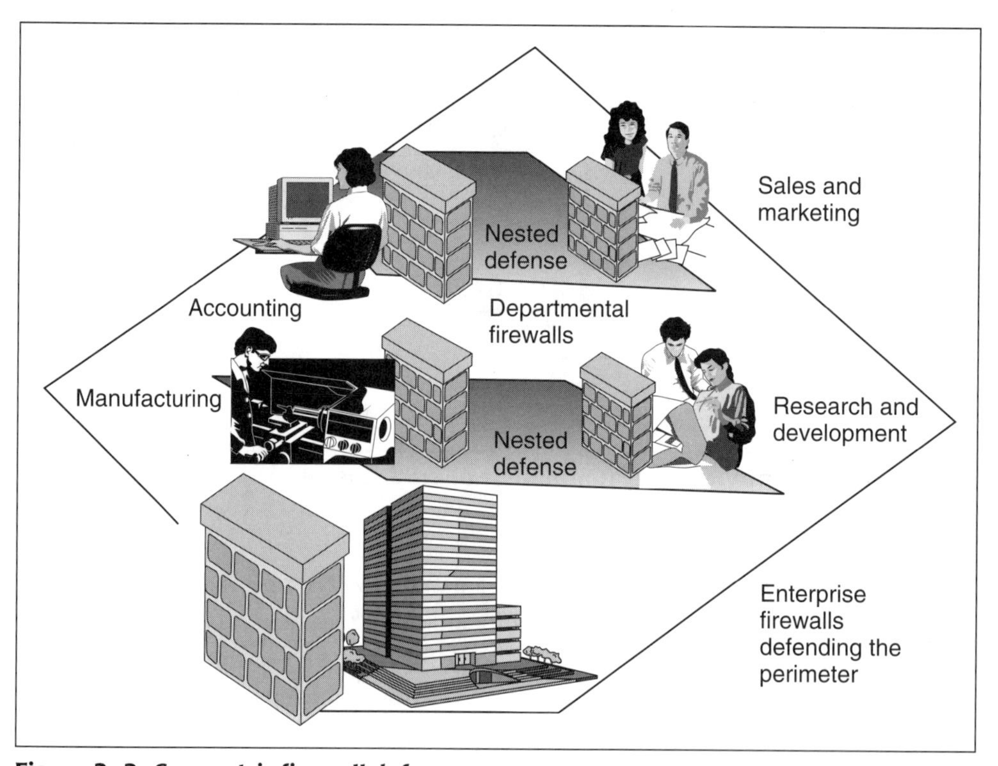

Figure 2–2 Concentric firewall defenses

In other words, firewalls, authentication servers, card keys, VPNs, and related technologies designed into traditional perimeter-defense schemes do not sufficiently mitigate risk to the e-business value chain but shift it from one area—mainly from the perimeter—to another. VPNs provide data protection in transit only, or from point to point. Unfortunately, data is still at risk before transmission and after it clears the security walls at the destination. To complicate matters, poorly chosen passwords, borrowed or confiscated card keys, and misconfigured security rules on network devices compromise access controls and user authentication. As such, if mini–perimeter camps are deployed throughout an extranet, for example, the network is vulnerable to internal breaches. These walled-off functional domains may also create performance degradation throughout the entire value chain, possibly supporting several enterprises. Finally, users perceive the barriers as being unnecessary and dysfunctional.

On the IT side, vigilance wanes as limited staff with limited resources are unable to cover the security demands of an open system, such as an extranet, e-tail store, or public Web site. When IT security staffs are frustrated owing to limited resources and inflexibility of the security infrastructure and end users are frustrated owing to performance degradation and compromised logins, security safeguards are subverted, and breakdowns in protection are inevitable. Clearly, what is needed for secure e-business is an adaptive security infrastructure that is scalable, fully functional without degradation of network performance, and centralized manageability with limited resources.

Principles of E-Security

E-security, or security for open information systems, allows access to information assets by outsiders. It extends, enhances, and complements the security measures provided by point solutions, regardless of whether the perimeter is known or whether data is in transit. E-security is also flexible and scalable and can be administered throughout the e-business value chain. The following principles of e-security recommend that you

- Understand the operational characteristics of your networks and the business objectives that are supported. Prioritize the critical importance of each information system and protect it accordingly.
- Refortify point security solutions, such as firewalls, authentication systems, and VPNs, with adaptive technology that maximizes effectiveness and limits premature obsolescence.

- Develop and deploy a comprehensive enterprisewide security policy, test it regularly for compliance, and update it when necessary.
- Purchase infrastructure products, such as intrusion detection systems (IDS) and assessment tools. Do your research! Consult an independent source to determine which IDS and assessment tools are best for your environment and support overall e-business security performance.
- Keep e-security simple. Institute simple policies and authentication systems that enable hassle-free access. Be diligent in achieving a security infrastructure that is easy to scale and manage on an ongoing basis.

The preceding principles suggest that e-security should be approached as a system life-cycle process supported by an adaptive security infrastructure. Only when e-security is such an ongoing, holistic process will you be able to minimize security risks and consistently achieve a satisfactory level of security in your e-business application.

Risk Management in the New Economy

Ironically, one of the key functions of e-security—and the success of e-business is directly dependent on it—is an established, physical-world business practice called risk management. From security's perspective, the greatest challenge for e-business entrepreneurs is making the connection between risk management and protecting the assets of online supply chains. Until recently, online business concerns were not making this critical connection, which is even more ironic given the openness of the Internet and online supply chains. Protecting valuable assets while also opening them up to online supply chains gave birth to the concept of e-security when executive management raised serious concerns about the risks of creating extranets and Web presence to pursue major e-business opportunity.

E-business managers are concerned with not only the risks to online assets but also the potential for economic risks when Internet-oriented supply chains are compromised. As the E-Trade example in Chapter 1 showed, the economic impact can be considerable. Consequently, managers in the e-business world are pinning their hopes and dreams for success in the old-world business practice of risk management. The concept of risk management inherently signifies an ongoing process whose time has come to ensure secure e-business operations. (See Chapter 13 for a complete overview of risk management.)

Organizations pursuing e-business opportunities ultimately become more open, especially as more and more trading partners are added. When potentially heterogeneous systems of distinctively separate enterprises are tasked to work as one system—an Internet supply chain—risk management becomes the critical success factor for e-security. But what exactly does risk management entail?

When an extranet, for example, is implemented between your enterprise and trading partners, the resulting digital value chain may consist of information assets generated by the software applications developed by each individual trading partner. Assuming that the applications were developed with Internet-enabled technologies, threats, related security risks, and vulnerabilities become more difficult to manage, especially when the digital value chain is modified to accommodate increased traffic from existing or new trading partners. The principles of e-security demand that threats, security risk, and vulnerabilities should be dealt with not only in your network but also in the networks of trading partners.

For example, approximately 130,000 ports, or doors, are possible in network devices and applications, although most systems have only a few hundred available, by default. Each door, if not adequately protected, presents a *window* of opportunity for an intruder to slip into either your network or that of your trading partner(s). Also, network applications are developed with certain application software systems or tools. For a given Java software release, for example, a hacker could possibly devise a means to exploit a weakness in the resulting application to gain access to your network with a *Trojan horse* or other exploit. A Trojan horse is a malicious program concealed within another, innocuous-appearing program. When it is run, the Trojan horse secretly executes, usually without a user's knowledge.

Typically, such vulnerabilities may be corrected with new software releases, or *patches*. E-security requires due diligence for addressing vulnerabilities in the software development platform used to develop your extranet application. Because software systems are continually being enhanced and upgraded, it's imperative that new releases and/or patches to software be implemented whenever they become available.

An exploited or compromised vulnerability is an attack on your system. Security threats could manifest themselves as a *port scan* by a potential intruder or an innocuous e-mail attachment. With so many windows of opportunities for hackers, your network should continually be monitored for vulnerabilities or successful system compromises throughout the entire e-business application system. In the open society of the Internet, hackers are always working to exploit weaknesses in your system. To complicate matters, disgruntled employees turned saboteurs are threats from within the enterprise. Thus, to successfully manage risk to the information assets and intellectual property of your Internet value chain, you must be vigilant and implement a comprehensive risk management process. This crucial framework of the process consists of the following functions/activities:

- *Policy management.* Are your security policies in compliance? Are they in synch with those of your trading partners such that they function as a unified whole throughout the e-business value chain? If they are working together, how often are *fire drills* conducted to measure effectiveness?
- *Vulnerability management.* Vulnerabilities can be classified by applications, devices, operating systems, development tools, communication protocols, and interfaces. What tools do your enterprise and/or trading partners use to detect vulnerabilities in network applications and infrastructure?
- *Threat management.* Due diligence with the first two activities minimizes threats overall. A key step in threat management is determining the potential number of threat sources. However, intrusion detection systems can be used to deter threats to the system, for optimal protection.
- *Enterprise risk management and decision support.* Coordinating security with your business partners is a best practice and a recommended approach if information assets are to be effectively protected. If the connection between risk management and an effective deployment of e-business security measures is achieved, the resulting security architecture should also provide an effective decision support system. (See Chapter 13 for an overview of risk management.)

The preceding functions represent *new dogs of war* for the IT security team. To make the connection between risk management and protecting information assets, it is important to institute and regularly perform those functions. These innovative processes are the sentinels of open information systems and the critical risk management functions for achieving e-business security. These activities must also interact with and support point solutions to create a cycle of continuous security protection on multiple levels. The end result is a system that responds quickly to security violations, monitors network security performance, detects vulnerabilities, mitigates risks, analyzes attack patterns, enforces policies, and implements strategies for prevention and survival.

As you can see, e-business requires a more sophisticated model than the perimeter-based security systems used by private or virtually private networks. E-business applications require that security be implemented and supported on multiple layers in the network. They also require management buy-in to enforce controls and to make the necessary investments in security measures. In actuality, e-business security depends on opportunity management as much as it does on risk management. Opportunity management consists of establishing and focusing on the objectives of the e-business value chain and implementing the requisite strategies, including a security strategy, to achieve related goals. If customers, partners, and internal users perceive that your e-business application is available and that steps have been instituted to attain a reasonable level of privacy, confidentiality, and integrity, you increase your chances of meeting e-business objectives. If implemented correctly and risks sufficiently mitigated, your e-business system can even monitor and reveal buying patterns and understanding of targeted markets, streamline operations, reduce related costs, and increase customer satisfaction. In other words, e-security enables e-business.

How E-Security Enables E-Business

From a business-to-consumer perspective, a site that is perceived to be safe and convenient builds brand loyalty and a critical mass of repeat customers who generate far more profit than do first-time customers. In an article in the September 1990 *Harvard Business Review*, Frederick F. Reichheld revealed that in a number of industries, repeat customers generate 200 percent more profit in the third year compared to the first and 600 percent more profit in the seventh year for a business.

From a business-to-business standpoint, customers, suppliers, partners, and employees perceive that the online supply chain is a safe channel to connect information assets. Consequently, revenues are increased and profits are higher when physical processes have been streamlined. Field personnel realize a competitive advantage from up-to-the-minute information, new opportunities are pursued almost at the speed of thought, preemptive strikes are achieved in established markets, and e-business can ultimately be pursued at the speed of information.

Specifically e-security enables e-business by

- Ensuring data availability, integrity, and privacy
- Identifying potential application, host, and network problems that could lead to interruptions in service owing to security breaches
- Creating secure, confidential channels for transferring and exchanging data along the entire online value chain

E-security enables e-business because the risks that compromise security are sufficiently managed to reduce or to eliminate economic losses and to minimize loss of business advantages if the e-business application is compromised. The bottom line is that e-security enables enterprises to pursue e-business opportunities virtually uninhibited, enabling as many applications for open access as necessary to meet e-business objectives. This is the e-security–e-business connection.

The E-Security Dilemma: Open Access versus Asset Protection

To underscore the significance of e-security, the model for e-security is designed to address the inherently conflicting goals of providing open access and the critical need for strict asset protection. At the core of e-security is a cost-effective architecture that permits low-level anonymous Web access with high levels of transactional security. At the other end, the greater the level of authentication and authorization control desired, the more comprehensive e-security's architecture must be. Another critical consideration is that e-security must factor in the use of the Internet for both internal network requirements and internal network requirements that involve supporting external activity, such as the DMZ. The level of security that is ultimately implemented varies according to the extent that the Internet is used for e-business.

In summary, the security model for networks that are private or virtually private emulated a security model of the physical world. Networks remained secure as long as access was controlled and requirements for authentication, data integrity, and privacy were met with reasonable measures. On the other hand, secure e-business depends on not only these requirements but also trust, confidentiality, and nonrepudiation. Also, authentication, data integrity, and privacy should be revitalized to ensure that the e-business opportunity is sufficiently protected to foster reasonable expectations. Therefore, the essential requirements for achieving e-security include trust, privacy, data integrity, authentication, nonrepudiation, and integration.

- *Trust* ensures that parties involved in developing, using, and administering the e-business application are trustworthy and that the supporting system and resulting transactions are free from events that cause suspicion, worry, and user concerns.
- *Authentication* verifies the identity of users, servers, devices, and systems and ensures that these network objects are genuine.

- *Data integrity* protects data from corruption, destruction, or unauthorized changes.
- *Privacy* keeps data private between authorized parties. Encryption is the underlying foundation for privacy and, perhaps, all e-security components.
- *Confidentiality* prevents the sharing of information to unauthorized entities.
- *Nonrepudiation* eliminates the ability to deny a valid transaction. Due diligence is required to achieve this critical e-business goal.

With these essential elements as guidelines, maximizing open access while minimizing risks to the enterprise's information assets can be achieved. Achieving an e-security balance also depends on the seamless integration of security across potentially heterogeneous networks. Without any one of these, effective e-security will become an elusive target at best. These and other concepts of e-security will be expanded further in subsequent chapters.

CHAPTER THREE

The Malicious Opponents of E-Business

Over the years, the list of establishments that have been hacked has included the White House, the U.S. Army and Navy, NASA, Ameritech, Bell South, Estee Lauder, Ford Motor Company, Hewlett Packard, Packard Bell, Microsoft, Amazon.com, eBay, E-Trade, and Yahoo, primarily by viruses and the denial-of-service (DoS) attacks. In the case of Microsoft, allegedly, theft and unauthorized modifications of source code intellectual property occurred.

Except for the White House, the Department of Defense, NASA, and Microsoft, the names are largely inconsequential because for every attack that is reported or uncovered by the media, hundreds more go unreported. The Department of Defense estimates that only 1 in 500 attacks is reported each year. Many others, especially banks, want to avoid the negative publicity and repercussions of such news becoming public. In a 2001 survey commissioned by the FBI's Computer Intrusion Squad and the Computer Security Institute, 91 percent of 538 organizations responding detected computer security breaches over the past year. Even more alarming, 40 percent reported penetration of their systems from the *outside.* This number grew in 2001 by 37.5 percent over the previous year. The survey also indicated that 186 of the 538 organizations, or the 35 percent willing and/or able, reported combined losses from security breaches costing approximately $377.8 million. Of these losses, $151.2 million resulted from theft of proprietary information; $93 million, to financial fraud.

In response to the juggernaut, the market for providing security solutions is expected to reach an estimated $700 million in 2002, up from $45 million in 1998. So the question remains, can hackers be stopped? Depending on whom you ask, the answer is a qualified no.

In this chapter, the covert world of the hacker is unveiled. Hackers are cocky; many are talented and remarkably organized. The pop culture side is reviewed in order to gain insight into the hacker's psychological makeup, along with understanding why a certain segment of their community holds such contempt for Microsoft. The potential controversy surrounding hackers and crackers is explored. Finally, how political forces are mobilizing in response to hackers and related implications is systematically laid out for your consideration. When you are done, perhaps you will be able to draw a conclusion as to whether the attackers are marauders, organized for ongoing and increasingly craftier incursions, or just cyberpunks seeking cheap thrills.

The Lure of Hacking

There is no nobility in hackers and what they do. You've seen the arguments. You have especially seen the side that argues that hackers ply their trade for an altruistic purpose or to benefit technology. Hackers hack for various reasons, including malicious intent and financial gains, but they hack mainly because they enjoy it! The traditional view is that hackers break in for the sake of improving security. There are even written accounts of a successful hack that was fully documented by the hacker and left behind for the benefit of the IT managers. An e-mail address was also included, for good measure, if the IT managers decided to engage in a constructive dialogue about the attack.

If you just spent hundreds of thousands of dollars rebuilding the root directory of your server or restoring one after it was totally wiped out, you will never be convinced that those feelings of loss, frustration, anger, and helplessness were experienced for some greater good of society. Ironically, those *crusader* hackers would most likely empathize with this and blame an attack of this nature on their colleagues known as *crackers.* The fact is, several categories of hackers exist, including the class originating from within your own networks, a group consisting of disgruntled or untrustworthy employees.

You have probably been exposed to the hype about why hackers hack. You have heard the reports and read the articles. But for the most part, hackers claim that they hack for the challenge or to impress their peers. No doubt some in this underground culture are trying to promote the cause for more secure networks, although their methods may be convoluted or their thinking a form of self-actualized reverse psychology. This is equivalent to a bank robber getting away clean with a robbery only to return to hand the money back to the teller with a message: "Next time, don't forget to add the exploding ink pellets, and trigger the silent alarm as soon as the perpetrator leaves. Also, add another surveillance camera over the side entrance." So the question is, does the end really justify the means?

Interestingly enough, a 1999 study commissioned by the U.S. military concurred that most hackers lack malicious intent. For the record, the study also revealed that hackers have an inherent interest in technology and are motivated by ideals. Because this appears to be the basis of the hacker psyche, another revelation suggested that hackers don't like the notion that information is private. The cultural *mantra* of the hacker underground is that "information wants to be free."

Moreover, although many hackers underachieve in school, they are still very competitive. They hack to be king of the cybermountain, to impress their peers, or to attract the attention of the press. The fact remains, there's nothing like the adrenaline rush of infiltrating that high-profile site to swipe classified information or inserting some absurdity, such as hyperlinks to porno sites in the Far East. The more successful the incursions, the more they hack. In the final analysis, this might be the best explanation for hacking: Hacking is simply fun.

Hackers versus Crackers

Do crackers start out as hackers and at some point cross the line, never to return to their original, malice-free mind-set? Or are crackers totally separate animals that are not on the same psychological continuum as hackers? Whatever the case may be, hackers, as a rule, don't like crackers. As far as hackers are concerned, crackers are the "bad-seed" perpetrators. The only reason that there is confusion between them is because of the media. Ultimately, crackers are the ones who are enticed to the "dark side" of hacking, when unscrupulous businesspeople pay them small fortunes to hack into the networks of competitors to steal business secrets or confidential information assets. In contrast, hackers are the ones who end up in corporate cubicles, turning their years of illicit practice into healthy salaries as system administrators or security specialists. After all is said and done, from the layperson's perspective, the boundary between hacker and cracker remains blurry at best. No matter whether it's a hacker or a cracker who breaches your network, it is still unauthorized access to your system.

In a *U.S. News and World Report* interview, a leader of one of the nation's top security teams stated that, as of any given time, hacker techniques tend to be 2 to 3 months ahead of the latest security measures available. The reason is that most of the skilled hackers literally work around the clock to exploit network security holes and, in the process, stay ahead of software fixes or security countermeasures. Typically, thwarted hackers are average to above-average hackers, such as middle and high school kids, who hack primarily during the summer months, out of boredom.

It is basically agreed that the hacker elite, or the top 1 percent to 2 percent, will probably never be thwarted or caught. Although the leader of one of the top security firms believes that it is possible to achieve 98 percent to 99 percent effectiveness, he acknowledges that the top two or three hackers in the world can circumvent any of the world's leading security measures. However, those close to the computer underground believe that security firms are distressingly behind the curve. They maintain that the *black hats* are just too cunning, too elusive, and too organized to be consistently deterred from mainstream security measures.

Hacker Groups

Cult of the Dead Cow

One well-organized hacker organization is the Cult of the Dead Cow, known for the development of Back Orifice and Back Orifice 2000, designed mainly to expose Microsoft's security weaknesses. As the name implies, Back Orifice is a GUI-driven software utility that enables unauthorized users to gain remote access to computer systems through the back doors of PCs running Microsoft's Windows 95 or 98. The Back Orifice incursion arrives at the unsuspecting machines as an e-mail attachment. Once installed, the incursion command structure gives the unauthorized user more control of the compromised computer than the user has at the keyboard. How's this for the *free* society of the Internet?

The Back Orifice program is available *free of charge* from the Cult's Web site. More galling, Back Orifice was released in Las Vegas in 1998 at DefCon, the largest annual security conference. Moreover, Back Orifice 2000 is already available. Don't forget to watch your friendly Cult Web site for any announcements concerning Back Orifice 2002.

Global Hell

The name couldn't be more apocalyptic! Global Hell debuted in the mainstream when it hacked into the White House Web site. In May 1999, the White House staff was confronted with a picture of flowered panties on its home page. Global Hell took credit for defacing the site for the whole world to see. Though flustered, Joe Lockhart, the White House press secretary at that time, was unwavering in asserting that Global Hell would think this less of a sport when the authorities caught the group. The FBI was brought in to launch what amounted to little more then a knee-jerk response. The FBI executed an 11-city sweep of 20 suspected crackers, but to no avail. (Crackers are believed to be the dark sheep, or the malicious perpetrators, of hacker culture.) The FBI seized the computers of the alleged perpetrators, but many industry observers were not impressed with the FBI's grandstanding. Because crackers are using nearly unbreakable, or strong, encryption, the FBI was relegated to grasping for straws, because little evidence was obtained after the raid. As an interesting epilogue to the story, one of the best crackers arrested in the raid was released after admitting that he had access to servers in 14 countries. The FBI retained his computer, which he promptly replaced. Guess what he's doing now? This former hacker, a high school dropout, has a legitimate job providing remote security for an Internet service provider (ISP) based in Denver, Colorado.

Script Kiddies

For the most part, hackers maintain that they trespass on systems strictly for the challenge. They insist that the holes they break through are routinely patched up and the systems administrator of the compromised system thoughtfully notified of the group's exploit and subsequent fix. The innocent-sounding Script Kiddies group, by contrast, is one of the most dangerous and malicious cracker groups wreaking havoc on the Internet. Script Kiddies revel in breaking up things whether by accident or on purpose, using tools typically downloaded off the Internet instead of programming the hacking tools themselves. For example, two of their disciples, California teens who go by the handles Makaveli and TooShort, literally ransacked a group of high-level military servers in 1998. In contrast to the *altruistic* hackers, their goal was not to enter and patch up but to enter and tear up. For the most part, Script Kiddies qualify as crackers, and their modus operandi is destruction.

Hacking for Girlies

Hacking for Girlies (HFG) are the political activists of the underground hacker culture. Apparently, one member of the group has been held in custody by the federal authorities since 1995. In September 1998, HFG hacked into the *New York Times*'s Web site in protest and to show discontent with a *Times* reporter who wrote a book that chronicled the comrade's capture. HFG referred to the reporter as a clueless moron. The *Times*'s home page was plastered with slogans demanding the release of the fallen HFG comrade. HFG has also hacked into NASA's Jet Propulsion Laboratory and Motorola in support of its incarcerated colleague.

Why Hackers Love to Target Microsoft

The superworms Nimda, Code Red, and Love Bug were designed to exploit vulnerabilities in Microsoft's flagship products. Code Red, which cost the global community $2.6 billion, affects vulnerabilities found in Index Server and the Windows 2000 index service. Nimda infected close to 2.5 million servers and users in less than 24 hours.

But it was the Love Bug that had by far the most costly impact on the Internet community. When the Love Bug was unleashed, the program writer, allegedly a Philippine student, exploited a vulnerability in Microsoft's Outlook e-mail program. Outlook automatically executes files with the file extension .VBS, for Visual Basic Script. A VBS file is a program that is used to perform certain routine functions for application programs. When the Love Bug arrived at the e-mail directories of the world's organizations—arriving first in Asia, then Europe, and finally the United States—it came as an innocuously appearing e-mail attachment with the file name love-Letter-For-You.TXT.vbs. The message in the e-mail prompted the user to "kindly read the love letter from me." As soon as the attachment was launched, the Love Bug did its thing, killing multimedia files with extensions .jpeg and .mp3, photo files, and other small programs. At the same time, it mailed a copy of itself to every user listed in the user's address book.

The ability to forward, or self-propel itself to others, qualifies the Love Bug as a class of virus called a *worm*. In contrast, piggybacking, or being attached to a document or certain files, spreads most other viruses. Once it's in a host, a worm usually affects just the host and the host's files. Infection is typically spread to other hosts by contaminated transmissions or the transporting of infected files by floppy disk.

The Love Bug's offspring, New Love (Herbie), considered even more vicious because it systematically deletes all files in the affected machine, thereby killing the host, also favored the beleaguered Outlook. For some time now, Microsoft has been getting its share of criticism about the security holes in its application products. The company has also received a lot of negative press about certain other products, such as NT. The problems with NT are so widespread that entire

Web sites are maintained just to disseminate information on software fixes developed by the NT user community. An excellent Web site devoted to this end is NT BugTraq, at *WWW.NTBugTraq.com.*

Many IT professionals believe that Windows NT is slow, buggy, and untrustworthy. And the new operating systems—Windows 2000, XP, and Me—are starting to receive their share of criticism, too. Similarly, security professionals deride Microsoft's operating systems because of how hackers favor attacking software systems believed to be porous, unreliable, and likely to crash often. Such drawbacks leave systems subject to attacks. The bottom line is that NT-centric networks with Windows 95 and 98 clients have gained notoriety for being inherently insecure.

In response, Microsoft maintains that security issues are an industrywide problem. The company is right to a certain degree, but when you own 70 percent to 90 percent of the PC software market, depending on the product, somehow this position isn't too convincing. Nevertheless, Microsoft promised an Outlook fix that reportedly would inhibit the program from automatically launching .VBS documents. By now, you will probably have the fix. If not, visit the Microsoft web site for details on how to obtain it.

Microsoft's feature-laden products deliver truly marvelous functionality but unfortunately have numerous security holes. Consequently, hackers love Microsoft because of the various classes of vulnerabilities presented by its suite of products. If any company can address these issues, Microsoft can because of its financial muscle. For now, however, we must learn to live with security threats, much like we live with the prospects of catching the flu. In the meantime, do what you can to protect your network before the next cyberflu season hits.

Meeting the Hacker Threat

As suggested previously, hacking is a full-time occupation. As one generation graduates, another appears to be eager and capable of fulfilling their mantra: Information wants to be free. To that end, they are remarkably organized with underground trade magazines, such as *Phrack* and *2600 the Hacker Quarterly,* and plenty of Web sites, such as Attrition, Help Net Security, and AntiOnline. They are even taking up with mainstream security conventions, as the Cult of the Dead Cow did to introduce its back-door hacking tool, Back Orifice. Many of the tools that hackers use are downloaded from the Internet, complete with instructions and without much difficulty.

For example, in 1997, the National Security Agency (NSA) conducted an information-warfare game classified as Eligible Receiver. The secret war game was initiated to test several scenarios. For one scenario, which turned out to be the most dramatic, the NSA downloaded the hacking tools directly from the Internet and learned how to use them there, as well. Consequently, the NSA used the tools to penetrate the Department of Defense's computers. Although it was a test scenario, the NSA hacked into a classified network that supported the military's message systems. Once inside, the NSA could have intercepted, deleted, or modified any and all messages traversing the network. The scariest aspect of this test was that the NSA could have denied the Pentagon the ability to deploy forces. If these types of tools were available several years ago, you can imagine what's available now. No wonder the federal authorities are up in arms. But realistically, what can they do? And more important, what are they doing about the hacker threat?

The federal government was responsible for creating the Internet, so it seems fitting that the government would ultimately try to police the Internet, although it is now maintained by private-sector concerns. Contrary to the popular notion that some hackers are beyond the considerable reach of the federal authorities, some very good hackers are serving time. (In December 1999, a 31-year-old New Jersey programmer pled guilty to causing the outbreak of the Melissa virus. The Melissa virus, unleashed in March 1999, caused at least $80 million in damages.) However, the federal agencies that are wearing the white hats are not ashamed to acknowledge that they are playing catch-up for the most part. Nevertheless, you must give them brownie points for taking the proverbial bull by the horns.

National Infrastructure Protection Center

At the forefront of the charge is the FBI's National Infrastructure Protection Center (NIPC). The NIPC became the lead agency when President Clinton signed *Presidential Decision Directive 63* on May 22, 1998. The charter of the NIPC, sometimes referred to as the National *Critical* Infrastructure Protection Center, is to fend off hacker incursions, both foreign and domestic. The NIPC is staffed with more than 125 individuals from the FBI, other federal agencies, and industry, far short of the target of 243 agents. Moreover, in its first two years of operation, NIPC's caseload grew from 200 to more than 800. Cases ranged from vandalism of Web sites to potential theft of military secrets.

The NIPC, working with the Justice Department, took on the responsibility of investigating the Love Bug epidemic. The team quickly learned that the bug was launched from the Philippines; within days of the discovery, a suspect was taken into custody.

Furthermore, the NIPC was largely responsible for staving off the potentially more devastating virus New Love. Shortly after receiving news of that virus's existence, NIPC issued an all-points bulletin to the world. As a result of this warning and the diligence of the nation's government and commercial IT security teams, New Love was virtually halted in its tracks. No sooner had the information disseminated than the NIPC teams determined that New Love, unlike the Love Bug, had originated in the United States.

Central Intelligence Agency

As you would imagine, the CIA has its hands full with both physical terrorism and cyberterrorism from the touch of a keyboard. According to the CIA, at least a dozen countries, some hostile to the United States, are developing programs to attack other nations' information systems and computers that control critical industry computers and infrastructure. The CIA has estimated that the United States was cybertargeted by a foreign nation at least once.

If you think about it, you can understand the magnitude of the potential problem and the CIA's concern. Computers run financial networks, regulate the flow of oil and gas through pipelines, control water reservoirs and sewage treatment plants, power air traffic control systems, and drive telecommunications networks, emergency medical services, and utilities. A cyberterrorist capable of implanting the right virus or accessing systems through a back door, or vulnerability, could cause untold damage that could, potentially, take down an entire infrastructure. The most sobering aspect of the challenge facing the CIA is that if hackers are as effective as they are through regular financial means, imagine how powerful they can be if sponsored by the government of a nation.

Other White Hats

In addition to the NIPC and the CIA, several other federal and international organizations are joining in the fight. Among others, the FBI's Computer Intrusion squad is responsible for conducting surveys to determine the magnitude of the problem, thereby providing direction as to what should be addressed in the war against hackers. The Computer Intrusion squad often teams with other agencies, such as the Computer Security Institute, to conduct research into security breaches in federal and commercial organizations.

Other federal agencies in the fight include the NSA, which is obsessed with the tools, techniques, and technology that hackers use to ply their trade. To assess their effectiveness, tests are often conducted against federal agencies, such as NASA and the DoD, which tend to be the NSA's favorite testing grounds. In one series of exercises, the DoD discovered that 63 percent of the hacker incursions that were launched against it in the simulation went undetected.

Finally the Department of Justice and the National Security Council are also playing key roles in the war for control of the Internet. Stay tuned. It's only just beginning.

In the international arena, there is the Forum of Incident Response and Security Teams (FIRST), which is a coalition of international governments and private-sector organizations established to exchange information and to coordinate response activities to the growing threat around the globe. For more information or a list of international members, go to http://www.first.org/.

PART II

Protecting Information Assets in an Open Society

It belongs to everyone yet no one. That is the reality of *the network* of networks: that behemoth called the Internet. Hackers love the Internet, and their tenet—"Information wants to be free"—is a throwback to the beliefs the founders held when the DoD commissioned the Internet for development in the 1960s. When the ARPANET (Advanced Research Project Agency Network), the ancestor of the Internet, was finally switched on in 1969, the founders envisioned a high-speed network freely accessible to a community of users sharing data. Security was clearly of no concern in the *free-spirit* climate of the day. Besides, the creators felt that to institute security measures would hinder the free flow of information and ideas. So in a sense, hackers are keepers of the faith for the Internet's true calling.

On the other hand, businesses did not come to the Internet as much for what it was originally as for what it had become. Ubiquity, scalability, and cost-effectiveness are the primary reasons the Internet attracted hundreds of thousands of businesses from around the world. More important, the Internet became a promise of exponential returns, alternative or supplemental revenue streams, new markets, and new, cost-effective channels of distribution in established markets.

Chapters 4 and 5 explore the divergent goals of e-business and the hacker community and why the war for cyberspace is taking on the *virtualscape* that it is today. Part II acknowledges the tremendous impact of intranets, extranets, and virtual private networks on the growth of the Internet and also explains why firewalls and VPNs were strategic in turning the first cyberwar in favor of mainstream organizations. However, the resilience of hackers in the face of point security measures and the need for businesses to regroup under more comprehensive security systems is also explored (Chapter 4). Chapter 5 explores inherent network complexities that limit the effectiveness of firewalls and VPNs, or point solutions, and expands on the risk management functions/activities introduced in Chapter 2. These steps are the critical elements required for instituting a life-cycle security process.

CHAPTER FOUR

A New Theater of Battle

Businesses are inherently risk averse. The riskier the opportunity, the greater the possibility that it will not be pursued. Given the risky nature of the Internet, it's amazing that its use has grown as much as it has and that projections for hypergrowth extend over the next 5 years and beyond. About 1995, when businesses began to incorporate the Internet for commerce, it was believed to be a much greater security risk than private networks. Thus, to enter the high-stakes game of the Internet, businesses began betting with intranets, extranets, and especially VPNs. In this chapter, the payoffs of such Internet-based business models are explored, as well as why they were so effective in winning the first war against hacker incursions. This chapter also explores the evolutionary impact of open access, its detrimental impact on point security solutions, and how the theater of battle for e-business opportunity was changed forever.

From the Demilitarized Zone and the Perimeter to Guerilla Warfare

When they began using the Internet to support certain business applications, not many businesses were willing to connect critical information assets to the Internet. In 1998, a survey sponsored by the Open Group, a consortium of global companies pushing for security standards on the Internet, revealed that only one in seven companies was willing to link critical applications to the Information Superhighway.

However, decision makers knew that they were witnessing something compelling in the Internet. It was everywhere: ubiquitous; it was free: the only costs for ISP access fees; and it could support however much or little of business applications as required: scalable.

The technologies of the Internet also were exciting and would provide an excellent medium for presenting business ideas in a manner that would impress employees, clients, and business partners. To get into the game, enterprises began incorporating firewalls to defend their networks on the perimeter from the outside world. Press releases, marketing information, or other enterprise propaganda were converted over to HTML or Java-based applications, which formatted information into multimedia presentations, graphics and rich text formats in an information server and placed them into a *demilitarized zone* (DMZ) outside the firewall. Another name for information servers deployed in this manner is the *public* Web site. Similarly, FTP (File Transfer Protocol) sites were established to support large file transfers.

Other organizations were so enamored of the Internet technologies that entire networks were being built to create internal networks called intranets. In other words, intranets became miniaturized Internets for their respective enterprises. Web and FTP sites were still installed in the DMZ, but *internal* Web sites in the intranet were being used to disseminate critical information to internal users. When enterprises realized that by linking their intranets to those of business partners, suppliers, and customers for competitive advantage, extranets were born. Up to this point, intranets and extranets remained closed, fairly private networks. Extranet communications were typically handled by *public data networks*—distinguished from the public Internet—provided by such common carriers as UUNet or MCI WorldCom.

It wasn't long before remote or nomadic employees needed a means to access the enterprise's intranet and the extranet of a partner, customer, or supplier. The logical step was to use the ubiquitous, cost-effective, and scalable Internet to enable access into the enterprise's intranet or extranet. Besides, firewalls and public data networks that provided the main communications backbones protected the core infrastructures. So networks were relatively safe.

Ironically, instead of being fended off by firewalls and private networks, hackers began to reassert their will and to recover ground that was lost from perimeter defenses and private communications. As remote employees, nomadic users, and branch offices of the enterprise, partners, suppliers or customers were gaining access through the Internet, plentiful Internet access points increased the opportunity for hacker incursions with tactics that resembled guerilla warfare. What was clearly needed was a way to allow remote and nomadic users to communicate safely over the Internet into corporate intranets and extranets. Hacker incursions were succeeding too often and, as more and more enterprise information assets were going online, stronger measures were needed to thwart hacker exploits.

When enterprises began incorporating the Internet as the primary communications medium for remote access, firewalls began to lose their luster as an enterprisewide defense. Early on, firewalls were thought to be a panacea because they controlled access to enterprise resources. Firewalls also took care of IP spoofing, one of the hacker's most effective gambits for gaining unauthorized access into enterprise networks. Using various methods, hackers would gain the IP address of a *trusted* network host when that host went out into the Internet cloud for Web surfing, e-mail, FTP transfers, and so on. After confiscating the IP address of a trusted host, hackers could use it to gain access through the enterprise's router and/or gateway, even though network access was gained through an *external* port (see Figure 4–1).

As long as the IP address appeared on the routing tables of the router and the gateway, the network devices didn't care whether a trusted—internal—IP address would suspiciously engage the network from an external port. Firewalls were able to negate IP spoofing by simply including a *rule* that stated that any internal *source* IP address appearing on an external access port is rejected, or not granted access into the network (see Figure 4–2). The logic was that any IP packet attempting to gain access to the network from external ports should never have a source IP address of a trusted host from the internal network domain. In other words, if it's truly an external packet, the source IP address should rightfully originate from an external network domain.

In addition to IP spoofing protection, firewalls enhanced the effectiveness of IP spoofing by giving networks another advantageous security feature in network address translation (NAT). Through NAT, firewalls were able to prevent *invalid* or *secret* IP addresses of internal domains from going into the Internet cloud altogether. This was accomplished by assigning the invalid or secret IP addresses to a *valid* IP address, usually the firewall gateway or network router (see Figure 4–3), Thus, for communication destinations outside the network, NAT instantaneously translated, or changed, the invalid or secret IP address to the valid

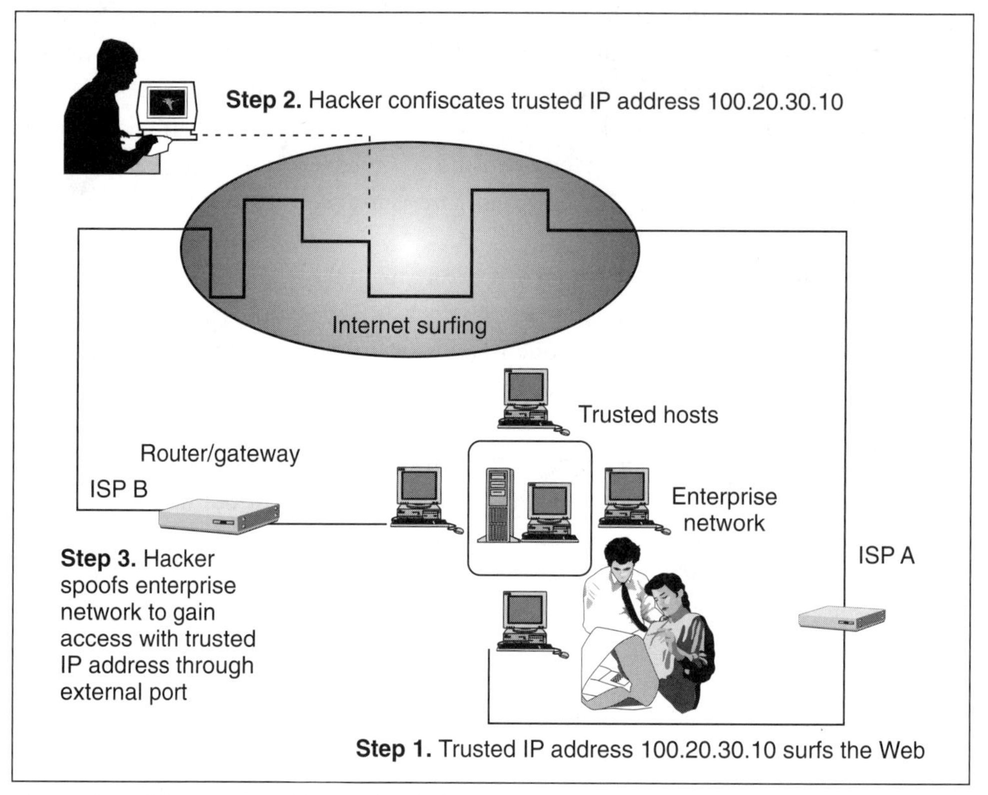

Figure 4–1 IP Spoofing, the hacker's favorite weapon

IP address of the firewall gateway or router when packets were queued for traversing the firewall port into the Internet cloud. NAT reverses the operation when receiving communication packets addressed to the valid IP address. After the valid or firewall gateway addresses of packets were translated or changed back to their original invalid or secret IP addresses, the packets were finally allowed to reach the destination of the host or hosts in question.

Another feature of firewalls is address hiding, which is similar to NAT, although address-hiding functions are usually provided by proxy-based firewalls. In contrast to NAT, proxy firewalls reconstruct packets at the proxy server or gateway and in the process substitute the IP address of the gateway for the source IP address of the original packet. This in effect hides the IP address of the trusted network host from external sources when traveling through the Internet.

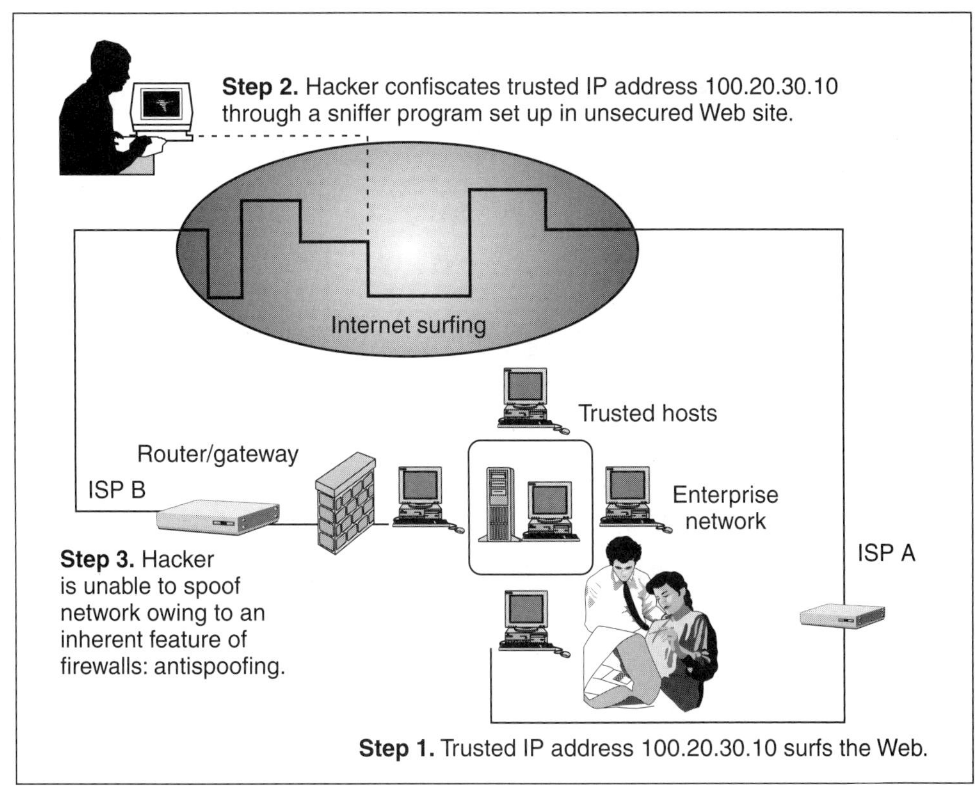

Figure 4–2 The firewall's antispoofing feature

Therefore, the IP addresses of trusted hosts of the internal domain could be invalid, secret, or valid.

Features such as NAT, address hiding, and anti–IP spoofing mechanisms worked in concert to create a powerful security solution for protecting the perimeter of an enterprise's network. However, when remote access was granted to mobile, home, and nomadic users, hackers were able to piggyback on these transmissions to gain access through the enterprise firewalls. With little or no difficulty, hackers easily read user IDs and passwords as remote packets made their way through unsecured network devices en route to enterprise networks. With all its powerful features, a firewall could literally do nothing to protect information transmitted via the Internet from remote sources. Unfortunately, in accommodating remote-access operations, the firewall lost its luster.

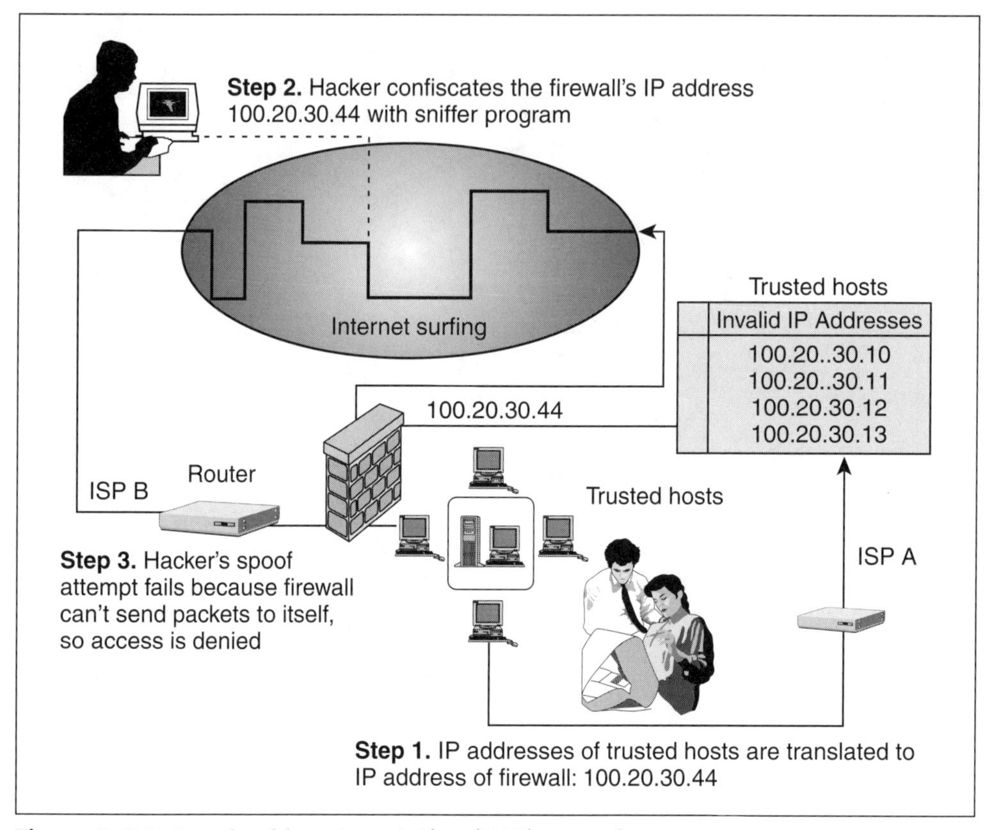

Figure 4–3 Network address translation (NAT) scenario

The Triumph of Intranets, Extranets, and Virtual Private Networks

The need was firmly established for remote users to successfully traverse the *virtual beachheads* provided by enterprise firewalls while protecting network infrastructures. The first cyberwar began when the business world decided to protect its interests. Intranets afforded the technologies of the Internet for the private use of enterprises while keeping networks private and hackers at bay. Extranets were a response to uncertainty, uncontrollable forces, and, especially, competitive pressures in business environments. Bringing suppliers, customers, and business partners directly online to enterprise computing assets was an innovation that marked an evolutionary shift toward a new business reality. However, the competitive advantages were soon being overshadowed when organizations turned to the Internet to connect remote and mobile users, as well as branch and home offices. These connections made good fiscal and business sense because of the economies of scale of the Internet, but they also provided windows of opportunities for hackers.

Hackers began using backdoor programs and *sniffers* to confiscate user IDs, passwords, credit card numbers, and other confidential information. Sniffers, often referred to as packet sniffers, are software utility programs that have become one of the favored tools in the hacker's arsenal. Hackers typically program and clandestinely install them in routers or servers of unsecured networks to do their covert bidding on the Internet. Hackers usually program them to monitor and/or search for data consisting of certain patterns, such as passwords, user IDs, or credit card numbers. Because a sender has no control over how a packet is transmitted through the Internet, data could be compromised any time and anywhere.

Also, viruses were being unleashed in unprecedented rates through e-mail attachments. The first denial-of-service ploys were being attempted using *ping of death.* Hackers discovered that by using huge ping packets instead of the customary size of about 64 bytes, a constant bombardment could deny access to legitimate users and eventually crash the site or, figuratively speaking, induce site death; hence, ping of death. With the Internet being used as the primary medium for remote access into corporate intranets and extranets, full-scale hacker incursions had returned with renewed purpose. It became obvious that firewalls alone could not protect enterprise networks.

Enter virtual private networks. The first VPNs were achieved by using simple encryption algorithms to provide a protective tunnel for information traversing the Internet cloud. Encryption scrambled strategic information while it was in transit between remote users and offices and the enterprise intranet or extranet. The first encryption systems, or VPNs, were by default proprietary. Examples include FWZ, developed by Checkpoint Software Technologies; SSH, by Data Fellows; and Simple Key Management for IP (SKIP), which was championed by Sun Microsystems.

VPN installations also incorporated strong, or two-factor, authentication to authorize users before tunnels ferried their data across Internet channels. Two-factor authentication systems work by giving users an object to possess, such as a token, and something to remember, such as a personal identification number (PIN). Strong authentication enabled VPN solutions with the most effective means of ensuring the identities of remote users. Typically, two-factor systems worked with an authentication server on the host network to authorize users or branch offices dialing in over the Internet. When authorization was successful, the encryption schemes would automatically take over to negotiate encryption between remote users and the host network. After successful negotiations, encrypted data transmissions would ensue, resulting in achieving virtually private network connections, resistant to hacker incursions. For example, sniffer programs, secretly installed in insecure systems along the Internet, were forced to read *ciphertext,* or scrambled text, instead of transmissions in clear text. For all intents and purposes, the white hats won Cyberwar I with the advent of VPNs, because intranets and extranets could ultimately operate as *virtually* private networks and protect remote access in the process.

VPNs are practical and becoming so commonplace that it would not be a stretch to consider them a commodity. The success of VPNs as a security measure for transmitting information through untrustworthy network clouds stems largely from the application and refinement of de facto and industry standards implemented as the underlying technologies. Tunneling protocols, strong authentication, digital signatures, and encryption work in tandem to create virtually private systems through public channels, such as the Internet.

One of the most important standards for VPNs is IP Security (IPSec). IPSec, *the* standard for VPN tunneling, was developed and championed by an Internet Engineering Task Force (IETF) working group of the same name. IPSec, Standards Track RFC (request for comment) 2401, was designed to administer security to the IP datagram, or packet level, by specifying a structure that would serve as the standard security format in VPN implementations (see Figure 4–4). Standardizing the structure for delivering security to IP packets was one of the major benefits of the IPSec initiatives. With IPSec as the standard for a tunneling protocol, vendors that incorporate IPSec into their VPN offerings ensure the interoperability of VPN solution components, regardless of the vendor.

Closely connected with IPSec is ISAKMP/IKE, the standard for key exchange, which deploys the encryption for data transmission on behalf of the tunneling protocol. In fact, ISAKMP/IKE (Internet Security Association and Key Management Protocol/Internet Key Exchange) is the default key exchange and management system for IPSec. ISAKMP/IKE, IETF's Standards Track RFC 2408, ensures that both end points of the VPN use identical keys for authenticating and encrypting IP packets. In other words, ISAKMP/IKE manages IPSec's encryption keys and enables deployment of security services to the data packet level. Services include negotiating the type of security to be delivered, authenticating the

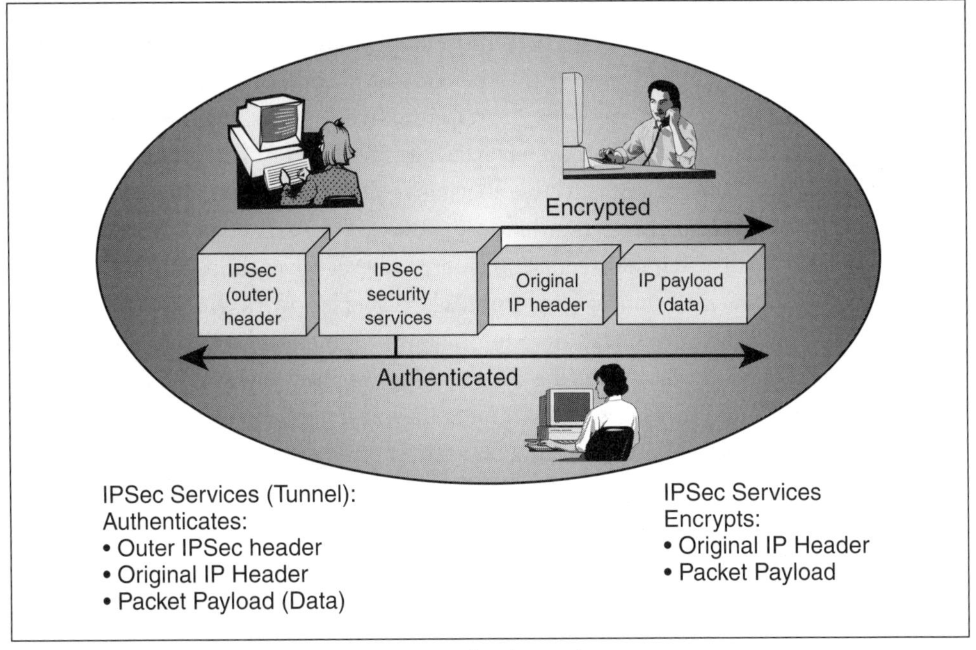

Figure 4–4 Security services of IP Security (IPSec)

tunnel, and encrypting the data transmissions. Finally, SKIP is an optional key management system supported by IPSec. In the long run, standards ensure interoperability, freedom of choice, and protection of investment.

The Vanishing World of Controlled, or Closed, Access

The Internet can be like hitching a ride on an 18-wheeler when *the path* to your expected destination is unknown or like a 747, with you as a VIP in control of your destiny and looking forward to a long-overdue vacation on the Riviera. When VPNs made their debut into the mainstream, businesspersons were finally on the plane, in control, and on their way to the Riviera: to paradise.

But before the beaches could get too crowded, others cut their stay short, and others found themselves back on the 18-wheeler again, chiding themselves for even thinking that paradise could be reached. As long as access was controlled and networks remained virtually private, or closed, businesspersons believed that networking paradise had been achieved when VPNs allowed them to harness the ubiquitous Internet as the enterprise's communications backbone. The ability to forgo the costs of leased lines, modem banks, and communications access servers resulted in cost savings of 40 percent to 70 percent. Whenever a mobile user or a remote or home office needed to be connected to the enterprise network, the individual in question was given the requisite strong authentication, VPN dial-up software, or an ISP account. Even if access was required from overseas, VPN solutions offered a more cost-effective alternative than leased-line solutions. In other words, VPNs were effective and indeed scalable. The archetypal network prototype had finally been realized. Or so enterprise executives thought.

VPNs were a boon to geographically dispersed work groups, WAN (wide area network) communications, and the bottom line. The ability to send private e-mail, access internal Web sites for graphical business literature, view executive video and audio messages, transfer large files, and participate in chat room–type discussions charged business climates with a new vigor for just-in-time information. The advantages that VPNs afforded businesses were unparalleled, especially working in conjunction with firewalls. A notoriously insecure medium in the Internet was being harnessed for all that it was worth. It seems you couldn't get too much of a good thing, because businesses were just getting started.

Uncertainty, dynamic forces, and especially the pace of the marketplace continually stimulated an already voracious appetite for extracting more from the Internet technology model. However, there remained some reluctance to take action. Perhaps mission-critical applications could also benefit from the just-in-time technology equation. Therefore, as a natural evolution in the e-business revolution, mission-critical applications were next to go online. The awesome significance of this meant that critical information assets were also going online by default. In fact, the prototypical archetype of the private network that was achieved with VPNs and firewalls is, for all practical purpose, gone forever.

The Impact of Open Access

Lessons were certainly learned over the years about general network security, such as managing user rights and instituting effective password policies and picture IDs for physical access. Such practices, coupled with such point solutions as VPNs and firewalls, should be sufficient security measures to protect the enterprise's information assets that made their debut on the Internet: not unlike lambs to the slaughter. When information assets went online, a virtual feeding frenzy was created indeed! Nomadic employees and remote users in branch and home offices found that in addition to accessing marketing collateral, technical white papers, press releases, product announcements, and the like, they were also able to check order status and product availability or to submit expense reports, sales forecasts, and capital requisitions online. Even more traditional work processes, such as accounting and human resources activities, went online, enabling enterprises to reduce operating costs by enabling accounting and HR staff to do their jobs from home by telecommuting. As a result, the costs of administration and expensive office space could be reduced considerably, further justifying the use of the Internet.

Not to be left out, trading partners, clients, and suppliers were experiencing the same trends at their end of the extranet. With such a flurry of activity, the point security provided by firewalls and VPNs was being stretched to the limit. To make matters worse, end user ad hoc requests were beginning to rear their ugly head.

An employee of company A, one of your extranet partners, needs to poke a hole through a firewall to accommodate a special UDP (User Datagram Protocol)-based application to complete research for a project deadline.

- Partner C just opened a larger than usual branch office, and so in addition to regular field-supported applications, the firewall and the VPN must allow access and privacy for connection into the enterprise's HR and accounting systems.
- Partner B just hired a company as a subcontractor for a major contract and must have access into the enterprise network for several of its satellite offices, but the security policy of the subcontractor is not as strict as the policy governing the original partner's extranet.
- Partner C reorganizes and the IT department loses several key managers to other responsibilities. In the meantime, the regular updates and revisions of the applications and operating systems on their part of the extranet have not been maintained.
- Partner A loses a major contract and staff must be reassigned or laid off, and the successful contractor extends job offers to others. The security policy is not adjusted to reflect this new development.

You get the idea. When systems are open and connected among trading partners and remote locations, you take on the "street-level" access of your partners, their risks, and those created by your own internal system.

The Correlation between Open Access and Asset Protection

When information assets went online, the opportunities for supporting remote activities, processes, and workflows, coupled with accommodating ad hoc situations arising from day-to-day business operations, were potentially staggering.

More important, at any time, where would one find the perimeter of the enterprise network? In the world of open access, it's a moving target at best. Furthermore, how do you keep the enterprise network safe with firewalls and VPNs when the perimeter is nebulous or when you can't keep up with everyone who is accessing your enterprise under various circumstances? What happens if your trading partner's regular program of system and software updates is deferred for a certain period? What is the impact to you when this occurs? Are network operations in compliance with the enterprise's security policy? Are you aware of every external individual or organization that your trading partners, customers, or suppliers have been granting access to for the information assets of the extranet? Are passwords and encryption keys being recycled effectively to reduce the potential for being compromised? Is the same diligence being followed for implementing strong authentication practices for temporary situations that support consultants or subcontractors? Is the security policy adjusted accordingly to reflect the start and finish of special projects? Are security audits performed periodically to assess the status of network security in general?

Even if a cross-company team were established with its own network operating center (NOC) to support the enterprise's extranet, keeping abreast of all the potential scenarios that may compromise security, especially armed with only point security measures, would be extremely difficult at best. Why? As has been shown, point solutions—consisting of firewalls and VPNs—are not inherently capable of protecting information assets in an open-access computing environment. As for the other questions, they will be answered over the course of this book. However, the most critical correlation between open access and protecting information assets is controlling *who* accesses the enterprise network. The effectiveness of e-security will depend on how well users are controlled through strong authentication measures. Strong authentication will take on a critical role in the level of success you achieve when protecting information assets in the new era.

The Role of Authentication and Privacy in the New Economy

The security benefits of privacy, data integrity, and confidentiality could never be achieved in an open system with VPNs and firewalls alone. More important, the trends that were catalysts for networks to function as closed systems are being vanquished by the migration to the open society of the Internet. A new business reality has ushered in an era of open access to critical information assets of global enterprises. Consequently, to protect critical information assets in open-access environments requires the implementation of a life-cycle or e-security solution rather than static point security measures.

The success of your e-security efforts is directly correlated with strong authentication and effective privacy. In fact, e-security and strong authentication are interdependent. In other words, in an environment that supports open access, the more control you require for user access, the more sophisticated the e-security infrastructure must be.

The importance of strong authentication was briefly discussed earlier in this chapter. Strong, or two-factor, authentication typically involves the deployment of smart cards and/or tokens. Two-factor authentication is a manifestation of the premise that authentication is most effective when users must *possess something,* such as the credit card–size token and *remember something,* typically a PIN number. As long as the PIN number is committed to memory and thus never written down, network access cannot be negotiated without a user's PIN, even if the token or the smart card is lost or stolen.

As effective as two-factor authentication is, it ultimately has some drawbacks.

- When *initially* assigned token and PIN, users may inadvertently expose numbers by leaving user IDs and PINs unattended.
- Users with multiple passwords may jot down their PINs temporarily, resulting in compromise by an unauthorized user.
- Tokens or smart cards can be lost or stolen.
- Disgruntled employees can compromise authentication procedures and access devices.

Any one of these scenarios compromises the security of the enterprise network, but it would be especially critical if the network supports open access. In closed environments, security managers may find that they are in a better position to discover a breach sooner. In contrast, real damage could be done in an open environment. This is why controlling who has access to information assets is so crucial. To enhance authentication efforts in open-access networks, security departments are also authenticating users through *digital certificates.*

Digital certificates provide yet another layer of security because a certificate authority (CA), such as VeriSign or Entrust authenticates user identities. Typically, a CA requires a user to produce proof of identity through a birth certificate and/or driver's license. Once an individual's identity is established, the CA signs the certificate and then issues it to the user. Digital certificates also contain encryption keys that are used in conjunction with VPNs or for encrypting passwords or PIN numbers in two-factor authentication systems. When it signs the certificate, a CA uses its *private* encryption key to initiate a process that places a digital signature on a user's *public* key.

The digital signature is an assurance that the user's encryption keys were not compromised before being placed into service. Signing the key also authenticates the user's identity because this action indicates that the user is ultimately who he or she claims to be. Using digital certificates with two-factor authentication systems gives security managers the greatest level of control over access to the enterprise's network in an open environment.

Finally, the challenge of controlling access when information assets are requisitioned for open environments is leading enterprises to single sign-on (SSO) and biometrics authentication systems. (SSO is discussed further in Chapter 10.) SSO authentication solutions are as much a response to streamlining end user sign-on procedures as they are to concerns about being burdened with too many user IDs and passwords. SSO systems allow IT departments to provide a user with one log-on ID and password for multiple systems, including client/server and legacy-based systems. SSO works in conjunction with digital certificate implementations and two-factor authentication to deliver the Rolls Royce of user authentication and log-on procedures.

Biometrics is growing in popularity and finding a place in the state-of-the-art for strong authentication. Biometrics relies on a physical characteristic, such as a thumbprint or a retina, to authenticate users. As the cost of such solutions decreases, enterprises will be implementing authentication schemes based on biometrics to protect assets in an open networking environment.

Summary

In today's Internet economy, markets are literally moving at the speed of information. To pursue opportunities successfully, companies have had to retrofit their information assets with information technology afterburners, such as J2EE, to support decisions, critical work processes, special projects, and regular business activities, responding to the unprecedented forces of a just-in-time business reality. But now that information assets are being repurposed for open access, are the gains worth it? In other words, are e-business opportunities worth the security risks that enterprises seem more willing to take than they were a couple of years ago? In addition, given the shortcomings of VPN and firewall security measures in an open-access environment, why is exponential growth expected for the Internet over the next several years?

All indications are that e-business will continue to forge ahead into the foreseeable future. In general, enterprises today appear to be more accepting of the Internet's potential security problems. The problems can be managed to mitigate their effects. Managing potential security problem effectively is even more critical when open access is granted to information assets. Saying that VPNs and firewalls should be enough just because you have invested in them is not sufficient. As long as risks are managed effectively, the potentially detrimental effects are minimized in the long run. E-security minimizes the security risks associated with open access to enterprise information assets.

This chapter focused on VPNs, firewalls, user authentication, and the shortcomings of point security measures relative to open access. Chapter 5 discusses the impact of certain network problems and other related issues in an open-access environment and expands on the tools required for implementing a managed, or life-cycle, e-security process. These tools will ultimately rearm IT departments with the weapons needed in the battle for information assets.

CHAPTER FIVE

Reempowering Information Technology in the New Arms Race

Integrating Web-enabled business processes with production business systems to allow open access is the new business reality. To implement an effective e-security program, IT departments must think out-of-the-box to forge all-new approaches to safeguarding the computing resources of the enterprise. Because point security solutions alone are not effective in protecting computing solutions of a new economy, IT departments, for all practical purposes, have no precedents to guide them. However, with creative approaches, management buy-in, end user support, and the emergence of intrusion detection, and vulnerability assessment tools, IT departments can fashion the security infrastructure, risk management, policy, and security procedures that are best suited to their organizations.

This chapter looks closely at the impact of tying in Web-enabled business processes with critical information assets of traditional production systems. It also discusses the importance of securing management and end user buy-in to ensure the success of a life-cycle security process. Finally, tools required to implement an e-security process are explored to rearm IT in an arms race with no definitive end in sight.

The Failings of the Old Paradigm

Whether you are providing an intranet for employees, an extranet for business partners, or an e-tail store for consumers, Web-enabled computing processes are changing the computing landscape forever. For example, through B2B, or extranet, solutions, business planning, manufacturing lead times, delivery schedules, and overall business cycles can be reduced considerably. This cuts operational labor costs, creates efficiencies, and reduces time to market. Fast and efficient companies will always have the competitive advantage and therefore a greater financial payoff, to the delight of upper management. This is the great allure of e-commerce. On the one hand, exciting returns on e-business investments have top managers lighting their victory stogies. On the other hand, as soon as they exit the smoke-filled rooms, concerns for security risks to the e-business channel are pondered aloud. Can we achieve business goals without Internet-enabled channels? Is our migration to e-business solutions keeping pace with appropriate security measures? Are the security risks identified and managed? Are all risks accounted for with certainty? Is the IT department equal to the challenge?

Protecting computing resources in a distributed processing environment, or protecting the *virtual mindscape,* has always posed certain challenges for IT departments. To alleviate the problem, IT managers embarked on *recentralization* efforts of distributed client/server operations. Successful centralization efforts fostered effective deployment of point solutions, such as firewalls and encrypted tunneling for point-to-point and remote dial-in communications (VPNs). However, when Web-enabled processes afforded open access to the enterprise's critical information resources, the security equation was fundamentally changed.

As discussed in Chapter 4, reliance on host-based access controls, such as passwords and IDs, or perhaps even strong authentication measures with perimeter defenses, quickly became unrealistic in the face of Web-enabled mission-critical systems. Today, IT managers are tasked with implementing a new comprehensive security model that accommodates risks, controls user access, and protects the infrastructure and the network's intellectual capital. However, IT managers' tallest order is to administer this security without disrupting the operational flow of the network while nonetheless preserving the integrity of sensitive information assets. A tall order indeed!

Unlike law enforcement agents, who always draw attention from passersby when apprehending suspects in public, IT managers must thwart intruders and attackers with as little attention and disruption to the network as conceivably possible. How will this be accomplished? As an IT manager charged with the reponsibility of building a working e-security model, you must first gain an understanding of the general operating and related areas that pose security threats to the enterprise network.

Infiltration of Rogue Applets

One of the most stirring concerns of open-access Web applications, as well as a true test to the effectiveness of e-security, is the insidious rogue applet. Applets are little programs or routines that are typically downloaded by your browser to execute on your computer. Applets perform client-side functions on your system in tandem with the server-side application that powers a particular Web site. Applets are typically developed with, for example, Java. ActiveX controls, which are similar in function to Java applets, must also be screened for malicious code.

Java applets are usually secure because of how Java operates within your system. Java cannot access a user's hard disk, file, or network system. In contrast, ActiveX controls are very different in *operation.* ActiveX controls can be developed in a variety of languages, including C, C++, Visual Basic, and Java. When downloaded from a Web application, ActiveX controls have full access to the Windows operating environment, making them a serious security risk. As implicated, ActiveX controls are limited to Windows environments only, whereas Java applets are platform independent.

Because of the compatibility of ActiveX controls with many *programming* environments, hackers use those controls to develop rogue, or malicious, routines. When they arrive, they can perform a variety of attacks, including zapping your hard disk, corrupting data, or setting up backdoor programs. Microsoft provides a fix of sorts that, when run on your system, allows the ActiveX controls to be authenticated before being downloaded.

Using ActiveX controls for Web application development is not as popular as using JavaScript for Web applications. In fact, although JavaScript is a full scripting language, it has become a favorite tool for hackers in *developing rogue code.* The bottom line is that you should be concerned about the potential of rogue applets and code in compromised Web sites and Web-based applications. If your end users have Windows-based systems, make sure that they have the necessary patches to handle rogue programs, especially those that are developed with ActiveX controls and JavaScript. The other option is to turn off JavaScript and ActiveX controls in your browser setting. Because Java-based applications can support ActiveX controls as well, you may also want to deactivate the browser's Java support.

Human Error and Omission

If you are responsible for security, controlling *human error and omission* will present the greatest challenge by far in protecting your network in an open-access environment. In the federal government, the "human factor" is *the* number-one security-related concern, according to a 2001 article in *Federal Computer Week.* To bring under control the various threats induced by human errors, you must evaluate the likely areas of problems or omissions to security and institute the appropriate measures. Human errors are most likely to cause vulnerabilities in deploying and configuring network devices and applications, user-access procedures and practices, and the application development process.

Configuring and Deploying Network Devices and Applications

Ill-configured network devices, applications, and security software cause one of the largest areas of vulnerabilities for network security. In Web servers, for instance, configuration errors are typically found in the Common Gateway Interface (CGI) programs. Among other things, CGI programs support interactivity, such as data collection and verification functionality. Too often, however, CGI programmers fail to account for the variety of ways CGI programming holes can be exploited. Hackers find CGI programming oversights relatively easy to locate, and they provide power and functionality on a par with Web server software. Hackers tend to misuse or to subvert CGI scripts to launch malicious attacks on the site, such as vandalizing Web pages, stealing credit card information, and setting up one of their most trusted weapons: backdoor programs. When a picture of Janet Reno was replaced with one of Adolf Hitler on the Justice Department's Web site, the investigation concluded that a CGI hole was the most likely avenue for the exploit. In general, demonstration CGI programs should always be removed from the production application before going online. The SANS Institute lists this problem in its top 20 most critical Internet security threats. (See Appendix A for a complete list.)

Misconfigured access control lists (ACLs) in both routers and firewalls create another class of security vulnerabilities in enterprise networks. In routers, human error in setup may lead to information leaks in certain protocols, including ICMP (Internet Control Message Protocol), IP, and NetBIOS (network basic input/output system). This category of breaches usually enables unauthorized access to services on DMZ servers. On the other hand, a misconfigured ACL in a firewall can lead to unauthorized access to internal systems directly or indirectly through the Web server in the DMZ.

Numerous other network components can create configuration error vulnerabilities. The preceding examples are among the areas that tend to be misconfigured most often. In general, when configuring network components, use checklists to ensure proper setup, test thoroughly for desired execution before components go into production, and *harden* your devices or applications. Hardening network devices involves eliminating or deactivating extraneous services, sample utilities, and programs that are no longer needed in the production environment.

User Access Procedures and Practices

One of the simplest problems in human error but potentially the greatest headache in providing effective security is poor password administration, which includes the use of weak or easy-to-guess passwords. (See the related discussion on strong authentication in Chapter 4.) The practice of using weak, easily guessed, and reused passwords creates one of the top categories of concern for network security vulnerabilities. This practically guarantees a means for compromising network servers.

Most enterprises have a policy for changing passwords at specified intervals. Surprisingly, in such environments, users tend to write passwords down on the last page of a desk blotter or a sheet of paper taped on the inside of a desk drawer or a file cabinet. Or if a password isn't written down in this manner, perhaps something easy to remember, such as the name of a loved one, pet, favorite color, or some combination, is used. These practices are not effective in preventing passwords from being uncovered and illegally confiscated. According to a June 14, 1999, *U.S. News Online* article titled "Can Hackers Be Stopped?" disgruntled employees and password thieves account for 65 percent of all internal security events.

Moreover, current research on the "human factor" reveals that an alarming amount of evidence indicates how frequently user passwords are learned by someone simply calling a user and posing as a system administrator. Even with stringent security policies governing the selection and administration of password use across the enterprise, this area of vulnerability is still a daunting challenge for security specialists.

To achieve effective password administration in e-business computing environments, enterprises are turning to awareness training for employees, coupled with incentives and/or penalties connected with acceptable practices for password use. Enterprises are also considering and implementing single sign-on (SSO) solutions that are offering enterprises a cost-effective solution, especially when multiple passwords are required. (See the discussion on strong authentication in the new economy in Chapter 4.)

Application Development/Tools

A growing area of concern for human-factor-driven vulnerabilities, and a potential Pandora's box, is application development practices and use of related development tools. Application or software development tools that are not properly maintained with patches or that are outdated or left in default configurations create one of the largest sources for security vulnerabilities. Applying patches and enhancements to development suites in a timely fashion is challenging enough, especially in IT departments with a high rate of either turnover or activities.

Beyond application development tools, however, concern is growing over the vulnerabilities resulting from the practices or techniques used when programmers write a given application. Prudence suggests instructing developers to write code that is free from common vulnerabilities and allowing *security* professionals to review that code for problems during development.

Reviewing code during the application process is not new. In fact, it is a throwback to the days when legacy applications were being developed for mainframe-based computing environments. However, code was and still is reviewed after application development is completed, usually during both pre- and posttesting stages, which may also include debugging activity. Code review, coupled with design and programming time, could be a lengthy process, depending on the size of the application. Even in the distributed processing world of fast application development cycles with GUI-driven development tools, code review can add 6 months to the cycle, which in today's fast-paced e-business world is seen as inhibiting to the business process.

Yet something must be done to eliminate common vulnerabilities arising from certain programming techniques. For example, nearly 40 percent of the common vulnerabilities and exposures listed in the Common Vulnerabilities and Exposures (CVE) database, an industry repository of various classes of software vulnerabilities, are buffer overflows. The CVE database is sponsored by The MITRE Corporation, a not-for-profit solution provider. If programmers are trained to avoid buffer overflows, such incidents can be reduced by 40 percent.

To address this critical area, one approach involves including security specialists throughout the volatile code-writing stage of application development to ensure that common security vulnerabilities are precluded from the final application. In effect, with the security and programming team working together throughout rather than after the entire cycle the potential risks are mitigated without slowing down development. Approaching an important security issue in this manner complements a life-cycle e-security program because hackers have fewer vulnerabilities to exploit over the long run.

Code review that involves security teams is a best practice and should be implemented for applications developed in-house. Unfortunately, however, this internal activity provides no remedy for vulnerabilities programmed in *vendor*-developed applications. One alternative is to request that an independent security code review be performed as a condition of purchasing the vendor application of interest. The fast pace of e-business markets—and most markets, for that matter—is an inhibitor to such stipulations being implemented on a regular basis, if at all. Owing to the serious implications of this issue, though, some vendors may incorporate a security review when writing their respective applications. In the meantime, until doing so becomes an industry practice, make certain that patches and upgrades are applied to vendor applications as soon as they are released. Applying patches on a regular basis will reduce the risks associated with vulnerability-exploited attacks.

For example, you might be aware of a security hole in the Remote Data Services (RDS) feature of Microsoft's Internet Information Server (IIS). IIS is deployed on most Web servers operating under Windows NT or 2000. Programming flaws in RDS are being exploited to gain administrator privileges to run remote commands for malicious intent. This vulnerability is listed on the SANS Institute's list of the top 20 most critical Internet security threats. If you haven't installed all the latest patches and upgrades to IIS, information on the fix can be obtained from the Microsoft site: www.microsoft.com/technet/security/bulletin/ms98–004.asp.

Ongoing Change in the Enterprise Network

Going for the "brass ring" in e-business is more than an acknowledgment of change; it is the ultimate realization that e-business channels are the logical, and perhaps even the only, choice for achieving business goals in a world that is continually changing. E-business channels are built on Internet technologies, which provide the flexibility, scalability, and adaptability enterprises need to thrive in dynamic marketplaces that are in an ongoing state of flux. That's the upside.

However, the downside is that the technologies used to build and to sustain e-business channels are by nature insecure. And when networks are continually being scaled, modified, or expanded to accommodate a just-in-time business model, those changes create another potential layer of security issues over and above inherent vulnerabilities. The areas that tend to change the most in response to e-business initiatives are remote-access points, internal host expansion, and autonomous operating departments.

Remote-Access Points

Connecting remote users into the enterprise network has always created distinct challenges for IT departments, even in private or virtually private networks. But as long as networks remained private, remote user connections could be controlled through a manageable, straightforward process because locations of remote perimeter points were known. In contrast, remote perimeter points in the e-business channel may or may not be known. (Refer to the section The Impact of Open Access in Chapter 4.)

For example, you may have *unknown* remote-access servers providing gateways into the enterprise network. Or, remote hosts may be connected by modem into an unauthorized remote-access server that circumvents the firewall and secure dial-in procedures, that is, VPNs. Or, *secured* remote users may be connecting to insecure networks of business partners. *Unsecured* remote users of business partners may be attempting access through remote control programs, such as Carbon Copy or pcAnywhere, connecting through an unauthorized remote-access server. These situations could compromise network security. However, the most common security exposures originate from unsecured and unmonitored remote-access points. Unprotected remote hosts are susceptible to viruses and incursions from backdoor programs, which can steal information and alter the contents of directories. Back Orifice is a classic example of a backdoor hacker tool that can completely take over the operation of remote-access points—hosts—by gaining administrator-level control.

Even secure remote-access points, where strong authentication and/or dial-up VPNs are in use, can provide vulnerabilities. For example, even though VPNs scramble all data involved in the transmissions, backdoor programs could piggyback into the remote host and find entry points into the host's directory or hard disk through unutilized services, such as SMTP (Simple Mail Transfer Protocol) or FTP (File Transfer Protocol). Effective handling of remote-access points in an open-access environment requires diligence, a robust security policy, and a firewall retrofitted with dynamic security surveillance capabilities, such as an intrusion detection system.

Internal Host Considerations

As e-business initiatives grow, internal host expansion, which includes workstations and servers, could present the broadest areas of concern for IT managers.

Workstation and server hosts are being added at any given time either incrementally or during department or corporate-level technology refreshment cycles. Many enterprises have a policy to refresh all enterprise servers and desktops every 2 to 3 years, on the average. Or, hosts are refreshed after every second generation of technology advancements. In between cycles, internal hosts are being added, moved, or changed to mirror the dynamics of e-business reality.

When hosts are deployed and go online, obsessive diligence is required to control potential security issues in open-access environments. If suppliers or customers are connected to the enterprise's network, you probably cannot do much to secure their networks. That responsibility lies with them, even with their support and buy-in to connect to your extranet. The effectiveness of the security measures they institute depends on skill level and, ultimately, the security policy that governs the rules of engagement from their perspective. You probably will also have no control of how often your business partners will add, move, and change network hosts in their environments. Hosts running unnecessary services, especially such services as FTP, SMTP, or DNS (domain name service), provide numerous avenues for intruders.

For example, the Berkeley Internet Name Domain (BIND) package is the most popular implementation of one of the Internet's ubiquitous protocols: DNS. DNS is the important utility that is used to locate domain names in the format www.xyz.com instead of numeric IP addresses. Without DNS, we would be forced to remember Web sites by numeric characters and periods. In a 1999 survey by the SANS Institute, 50 percent of all DNS servers connected to the Internet were found to be running vulnerable versions of BIND. Through a single vulnerability in BIND, hackers were able to launch attacks against hundreds of systems abroad. Attackers erased the system logs and gained root access to the hosts in question. In this situation, the fix was relatively simple. The BIND name utility was disabled on *unauthorized* DNS servers (hosts) and patched on *authorized* DNS servers. Only UNIX and Linux systems were affected. (See Appendix A for the top 20 vulnerabilities.)

Deployment of hosts may create other problems as well. System administrators may leave *excessive* file and directory access controls on NT- or UNIX-based servers or allow user or test accounts with *excessive* privileges. In both cases, excessive directory controls or user privileges provide vulnerabilities and exposure for exploitation. If these issues aren't addressed, they provide an ongoing level of security risk to the entire network every time a host goes online. In the meantime, if you are the sponsor of the extranet, you must implement the policies, procedures, and overall measures that protect the privacy and integrity of the entire computing environment. You may need to provide a gentle reminder to your counterparts in partner organizations about the potential security risks caused by excessive user privileges, unnecessary host services, and related problems as well.

Maverick Operating Departments

Perhaps it is a little unfair to classify this section as such. However, if you are an IT manager, you can relate to this category. Every IT manager has dealt with a department that is difficult to please or is never satisfied even when it gets what it asked for. The managers, typically impatient, will do things that other departments are supposed to do for them, such as hire staff members who should work for those departments. Independently operating departments are where you are likely to find, for example, the addition of an insecure NT-based Web server that the department has set up on its own. Such a department may have other systems that circumvent enterprise security policy and perhaps present security exposure to e-business applications.

In e-business environments, the corporate security policy should disallow applications that are developed by such departments to go online unless they are sanctioned with the appropriate security certifications. Getting independently operating departments to conform to enterprisewide security policy may be a daunting challenge, especially if the IT department's function is perceived to be decentralized or not fully supported by executive management. If you are confronted with a situation like this, the enterprise security policy will never be fully adhered to unless the IT department has the full support of executive management. Lack of accepted and officially sanctioned security policies, procedures, guidelines, and perhaps even a minimum baseline for standards is a significant underlying cause for vulnerabilities and exposure.

Deploying and Maintaining Complex Layered Client/Server Software

The Internet supports a dizzying array of computing software suites that can be combined into a maddening number of computing environments that operate under various flavors of UNIX, Linux, NT, and more recently, Windows 2000. The Internet also supports numerous protocols for communicating multimedia, EDI (electronic data interchange), plaintext, and encrypted data. More important, no matter what suite of software is used to develop your e-business applications, if it can be Internet enabled, communications can transpire among a huge variety of computing environments. However, there is both good news and bad news.

First, bad news: Many Internet technologies are insecure; it's simply a question of degree. For one thing, TCP/IP is inherently insecure and is the basis for all communications throughout the Internet. As for operating systems, NT and Windows 2000 have more security vulnerabilities than UNIX, but hackers are still able to achieve administrator or root access, respectively, through a variety of means. Of the Web development tools, JavaScript poses many more of the security issues than, say, J2EE or Java. (See the section Infiltration of Rogue Applets earlier in this chapter.) Although Web presentations created by these tools can be unbelievably spectacular, hackers can compromise Java- and JavaScript-based sites, especially if Java-based sites incorporate ActiveX controls. Applets resulting from these compromised systems can create gaping holes through network security.

The good news is that with life-cycle security measures, you can protect your e-business computing environment with its complement of client/server software layer by layer. But first, you need to appreciate that the potential for vulnerabilities exists on every operating layer of the e-business application (see Figure 5–1).

Contemplating the vast number of software systems potentially available for creating complex e-business applications and the *total* number of common vulnerabilities and exposures at any given time boggles the mind! For instance, the CVE database contains more than 1,600 entries. The CVE database initiative is a concerted effort to standardize vulnerability identification—names—thereby increasing interoperability among security tools. This initiative to create a common lexicon of vulnerabilities is supported by business, government, and other institutional concerns. (For more on CVE, see Chapter 11.)

Fortunately, the software systems and tools you are using to develop and to build your e-business channel are most likely a manageable number. The point is that, no matter how large the enterprise client/server software pool is, you need to have a complete accounting of the software titles, release levels, and/or version numbers that are supporting production environments. Just having this accountability is a critical step and will be instrumental in minimizing the impact of potential vulnerabilities.

Another important step is facing up to the reality that vulnerabilities exist throughout the operating layers of the application suite: from the operating system to the applications themselves (see Figure 5–1), especially in Internet-based technologies. Vulnerabilities also exist in *running* services. Running services are the protocols, utilities, subroutines, objects, and so on, that applications use, call, or require instructions from when running in production operating mode. Examples of running services that applications use or call include *automatic*

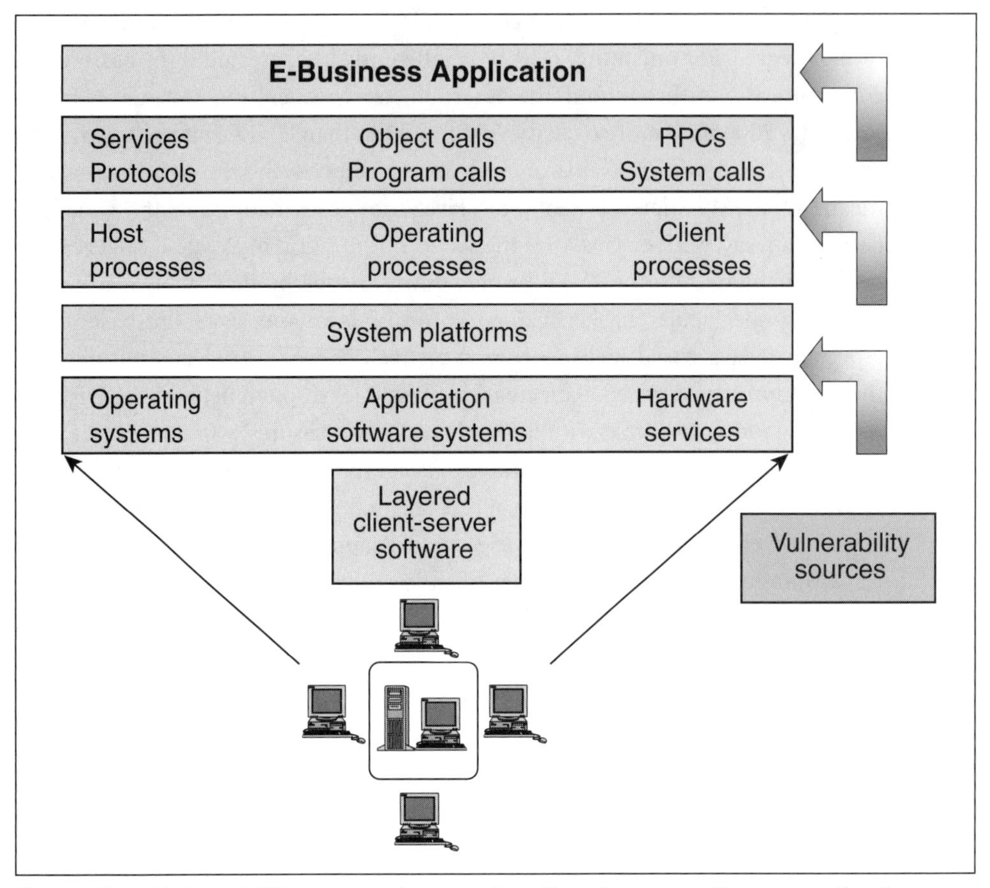

Figure 5–1 Vulnerability sources in complex client/server software applications

execution of .VBS (Visual Basic Script) files or JavaScript on the application level, NetBIOS functionality on the operating system level, or SMTP or FTP on the network communications level.

Perhaps the most-overlooked sources of vulnerabilities are default settings and passwords, demo accounts, and user guest accounts. When setting up complex systems, it is easy to miss simple things, such as addressing the issues of default settings. For example, SNMP (Simple Network Management Protocol) is a favorite utility of network administrators, who use it to monitor and to administer a variety of network-connected devices, such as routers, printers, and hosts. SNMP uses an unencrypted "community string" as its only authentication mechanism. The *default* community string used by the vast majority of SNMP devices is "public." Some vendors change the designation of the community string from "public" to "private." However, attackers use this vulnerability to reconfigure and/or to shut down devices remotely.

Another favorite default-setting target of hackers are administrator accounts with default passwords included with database management systems, for example, or default maintenance accounts of services in UNIX or NT with no passwords. In either scenario, attackers guess default passwords or access services not password protected, to gain access to root or administrative privileges in hosts, including those behind firewalls. Default settings are listed in the SANS Institute's list of top 20 vulnerabilities. (See Appendix A.)

Finally, in deployment of complex layered client/server software, flaws, bugs or other functional defects are a critical source of vulnerabilities that are popular targets exploited by hackers. To minimize the impact of flaws or bugs in the application programs, you should be diligent in applying patches and updates to the software. Often, patches are a response to security breaches after a discovery is made known to the software vendor by a user group, for example. Although a particular bug or a functional weakness may not have affected your network, installing the patches ensures that the bug or flaw in question will not affect your application in the future.

In summary, you have to address four classes, or categories, of vulnerabilities in complex client/server software installations: software bugs, misconfigured devices, default settings, and availability of unnecessary services. Figure 5–2 recaps these important classes of vulnerabilities in e-business environments.

Shortage of Human Capital

One of the greatest challenges for achieving effective e-security is using the professional talent to execute an effective life-cycle security program. In general, skilled IT professionals are in short supply, especially IT security professionals. The number of IT security personnel is much less than .1 percent of the total enterprise employee population. To put this in perspective, in organizations with 100,000 employees, it is not unusual to find fewer than 25 security specialists. In medium to small enterprises, the security-staffing ratio quickly diminishes to zero. An important branch, consisting of 22,000 individuals, under a critical federal agency has only one security specialist. Coincidentally, this branch has been hacked numerous times, including from China.

The main reason security expertise is lacking is directly attributable to years of neglect in both the academic and professional worlds, owing to the paucity of

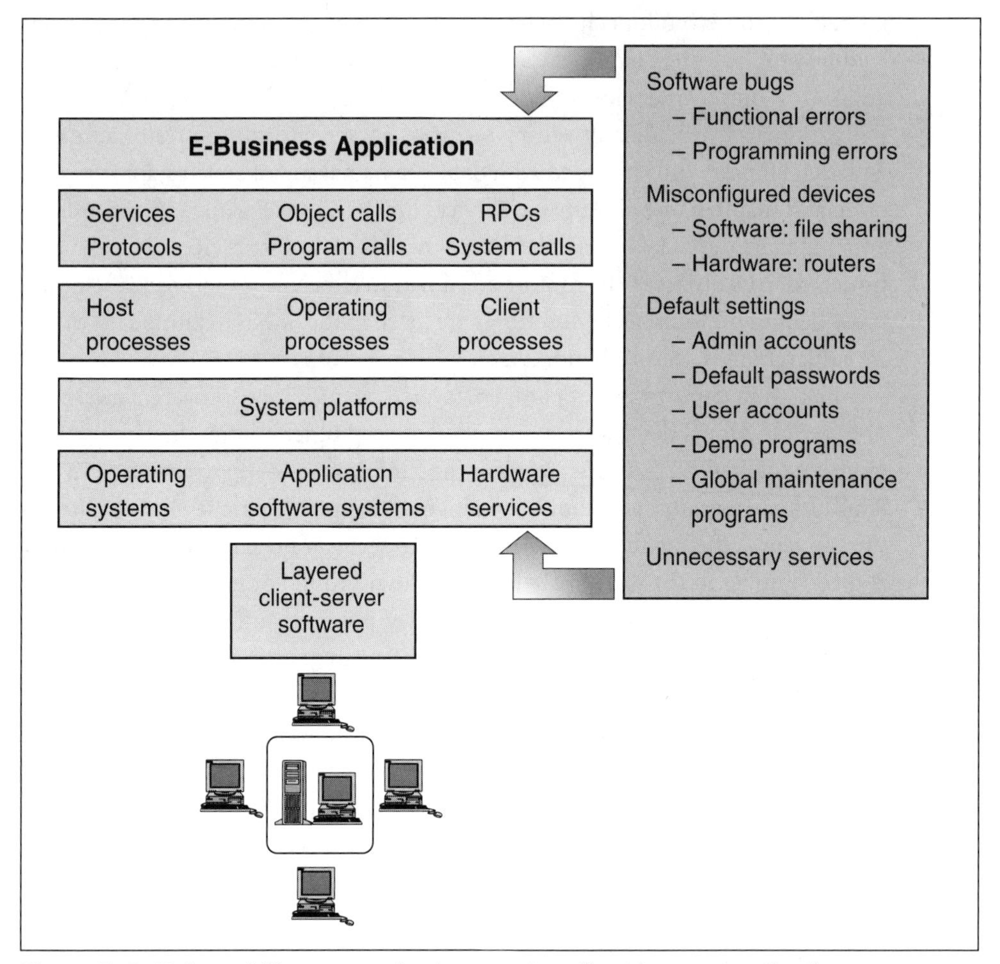

Figure 5–2 Vulnerability categories in complex client/server applications

education and training curriculum and programs. Another reason that e-security professionals are scarce is the perception, and it's only a perception, that cost-effective security is complex. IT managers in general believe that the ability to find security talent well versed across systems, operating environments, network protocols, middleware, Internet-based technologies, and applications is rare indeed. To exacerbate the problem, to identify security professionals who can work successfully with e-business managers is not unlike searching for a needle in a haystack.

To address this severe shortfall in security specialists, academic, business, and government institutions are developing and instituting security curriculums and programs. For example, Purdue, Idaho State, and James Madison universities are offering degree programs in IT security. In the meantime, security companies, such as Axent Technologies, Internet Security Systems, Cisco Systems, and NetSonar, are providing security tools that enable enterprises to achieve e-security goals by leveraging limited security staffs.

Rigidity of Enterprise Security Policy

An inflexible security policy can create vulnerabilities in the security infrastructure in the same way that a weak security policy would. However, a flexible security policy is difficult to achieve, particularly in an environment that is supporting open access to information assets. Developing a thorough and achievable security policy is a best practice, but if the enterprise perceives it to be too rigid, users will circumvent the guidelines in the name of pursuing business goals. For this reason, an exhaustive review of the enterprise's business processes should be completed to determine their alignment with business objectives. If, for example, a department participates in regular chat room discussions, the security policy should reflect this and allow AIM/ICQ (AOL Instant Messenger/"I seek you") communications through the firewall. But in doing so, it should also note when these chats take place. If the AIM/ICQ sessions take place every Tuesday and Thursday morning and you suddenly see a session happening at 9:00 AM on Saturday, this activity is suspicious at best. To eliminate any potential vulnerability, the security policy should not allow any AIM/ICQ sessions for the two hours before noon on Tuesday and Thursday. If more flexibility is desired, the other three days of the work week can also be designated to support additional newsgroup activity.

Building a security policy for one department, let alone an entire enterprise, can be an extensive, time-consuming process. Nevertheless, that process must be undertaken to achieve an agile security policy for an open-access computing environment. Otherwise, users will circumvent the security policy to pursue activities they believe are in the best interest of performing their duties. If the security policy is not robust enough to support these activities, the entire e-business environment is exposed to risk from security violations.

Tools for Rearming the IT Manager

The purpose of this book is to make you a general of e-security deployment, readiness, and resolution. The ability to wage an effective campaign of security against threats in the *wild*—a name that is growing in popularity for the Internet—and from within the enterprise will require a total refocusing, perhaps even a metamorphosis, of the enterprise: from end users to the CEO. Instituting a comprehensive life-cycle e-security program for the enterprise's e-business initiatives will generally require changing the way the entire enterprise regards network security. Making this happen is a monumental task but can be achieved with relatively simple but evolutionary guidelines.

Guidelines for E-Security

One guideline of critical importance involves convincing the entire user population that *everyone* is responsible for protecting the network. This will prove to be the most critical underlying task for effective e-security. Each employee, including the CEO, must become a security officer, someone who is directly responsible for the security risks that may be generated from his or her individual work activity. This, of course, cannot be achieved overnight, because employees must first gain an understanding of what those risks are or at least what behaviors threaten the security of the enterprise's network. At a minimum, therefore, a program of introductory, refresher, and job-specific safeguards should be implemented. In empowering each individual, you can be creative. Make sure, however, that costs for an enterprisewide training program are kept in proportion to the costs of your overall life-cycle security program.

Another guideline is deciding what level of exposure to risk is acceptable to the organization's computing environment. The answer could range from a zero tolerance to acceptance of some risk. For example, the concerns of a defense contractor bringing suppliers online to support classified projects might be more far-reaching than those required by an apparel manufacturer connecting with its suppliers. The defense contractor might decide that a *host-based* IDS that protects individual user hosts, as well as a network-based IDS and personal firewalls, would be needed. In contrast, the apparel manufacturer may determine that only a *network*-based IDS, along with a corporate firewall, would do the trick. In short, requirements for e-security will vary from enterprise to enterprise. Don't anticipate any uniformity of requirements, because there are no cookie-cutter e-security solutions.

Another crucial guideline for establishing comprehensive e-security is determining your network's perimeter. In open-access e-business environments, the network's perimeter is typically difficult, if not impossible, to discern. However, a concerted effort must be made to determine whether the network's perimeter can be ascertained with reasonable certainty. In B2B environments, the more nebulous the perimeter of the network, the stronger the authentication system should be. In other words, if you bring on several multinational suppliers, each supporting a variety of clients in a variety of markets, the more likely you will not know where the perimeter of your extranet will fall. You may start off with a connection to a certain office with certain individuals. After a consolidation, reorganization, merger, or other developments, for example, you may end up with another office managed by new individuals. Therefore, the potential for change may create a more nebulous or indistinct network perimeter.

Generally, the more nebulous your network's perimeter, the stronger the user authentication system you should use. A digital certificate server and/or biometrics are perhaps the strongest user authentication available for authorizing individual access. In other environments, a two-factor authentication system using smart cards with related access servers might be sufficient. Whatever the case may be, the greater the risks in controlling authorized access to your network, the stronger the user authentication should be.

To recap, the following guideposts are critical to the establishment of an effective lifecycle security program.

- E-security is the responsibility of all enterprise users.
- Determine what level of risk exposure is acceptable.
- The more nebulous the perimeter is—the more open the environment is—the stronger the user authentication should be.

Ensuring that these guidelines are followed is an iterative process. For instance, it may take several training or awareness sessions before users acquire the mindset and behavior that contribute to an effective e-security program. Furthermore, the greater the number of user IDs and passwords an individual requires to perform his or her job, the more resistance you will encounter before strong passwords are used consistently, especially if they are more difficult to remember and should be changed at regular intervals. Therefore, as you implement your e-security measures, make implementing these guidelines your mandate. At some point, you will consistently see use of strong passwords; know the security status of the network in terms of acceptable levels of risk and control authorized access to information assets even if the network's perimeter continually changes.

Make sure that you attain the results that you want, even if you don't realize them until after your security measures begin to be implemented. The strength of your e-security program depends on how closely you meet these critical guidelines. After they are sufficiently addressed, you can begin or continue to tackle the establishment of your enterprise's life-cycle security system.

Enterprise Security Policy

Implementing a life-cycle security program will probably be one of the most important IT projects of your career. If implemented properly, e-security will not only protect your network's perimeter and infrastructure but also enable e-business. (See the section How E-Security Enables E-Business in Chapter 2.)

In general, an e-security program is fashioned from the effective integration of five processes and their related tools: security policy management, risk management and assessment, vulnerability management, threat management, and attack-survival management. Note that as discussed in Chapter 2, an e-security program is essentially the risk management process for network security, coupled with attack-survival management. The e-security process encompasses the methodologies and tools that enable you to tailor a life-cycle security solution for your specific needs. Because of their critical importance, each of these five subject areas is discussed in a separate chapter. Therefore, this chapter will not go into any further detail. The important fact to remember here is that these processes are the building blocks of e-security. They provide the necessary tools to deploy precision countermeasures that are designed to thwart frontal and guerilla hacker incursions and internal saboteurs.

If you have built an intranet, extranet, B2C channel, or public server in the DMZ, most likely it was developed to conform to the security policy of the enterprise. The type, breadth, and effectiveness of your security measures are only as strong as the related security policy. The security policy defines who is authorized to access the enterprise's information assets and intellectual capital, along with the standards and guidelines about how much and what kinds of security measures are necessary and the procedures required to implement them.

In many organizations, the security policy is not a *living* document but rather an articulated understanding established by tradition and general business practices. In all circumstances, a *written* security policy is a best practice. In closed networks, an *unwritten* security policy hinders the setup and implementation of firewalls, VPNs, and authentication systems. An unwritten security policy also fosters circumvention by maverick employees and departments, limits consistency of understanding by the general populace, and makes it difficult, if not impossible, to enforce disciplinary measures when violations occur.

In open networks, such as an extranet or a public DMZ, an unwritten security policy increases the security risks to information assets, regardless of the security measures in place. For example, if left to personal interpretation or judgment, a public Web server in the DMZ might also serve as an FTP server. Because of the potential and inherent vulnerabilities in both systems, a Web server, especially an FTP server, should never be housed on the same *physical* server, especially when it resides outside the firewall. In general, FTP servers must keep ports open to accommodate potentially heavy file transfer requirements. (If you are using a *stateful* inspection firewall, you need to have only a single TCP port open for inbound traffic but must be able to have the FTP server establish multiple UDP connections for outbound traffic.) At a minimum, FTP servers are vulnerable to probing attacks, as well as to the use of unencrypted passwords, buffer overflows, and both the PORT and SITE commands.

An experienced network engineer would never put an FTP server together with other services. However, a less experienced engineer or financial constraints might make this a reality. A written security policy would prevent such security risks by clearly stating the requisite guideline(s) for a given situation and the resulting repercussion in the event of a violation. Running open-access computing environments without a written security policy is like playing Russian roulette indefinitely. Although the revolver has only one bullet, the weapon is still loaded, and it's going to get you.

Therefore, write down your security policy. Typically, a security policy should be no more than three to five pages and have a life of three or more years. Most important, the security policy should be written so that it is resilient to change. (See Chapter 10 for more on security policy development.) A written policy eliminates vulnerabilities to the security precautions that are instituted, facilitates the implementation of new security measures, increases the general understanding of security, and simplifies the ability to legislate disciplinary measures when violations occur. As you bring on business partners, your security policy should reflect the various types of information that they are allowed to access. Whatever diligence or security measures they must institute to connect to the extranet should be clearly delineated, complete with the recommended procedures about how the related measures are implemented. In general, consult with your business partners when penning the security policy, and give them a copy when the policy is completed.

If the enterprise already has a written policy, it may require revamping to sanction the e-business initiatives that executive management is pursuing or is planning to pursue. Even if it has been changed before the current life cycle has expired, the present security policy should be reengineered to accommodate e-business initiatives. Implementing the security policy in your enterprise will be challenging enough. But with your business partners on board and governed by the security policy, risk and exposure to your network are dramatically reduced from their networks, and you inherit a fighting chance in providing a safe computing environment for everyone.

Summary

Ideally, you now have a better appreciation of what you should address to establish enterprisewide e-security. Many of the deployment, computing, and related business activities that create vulnerabilities in the enterprise network can be managed before any specialized security measures are purchased. Controlling several key sources of vulnerabilities by applying patches, hardening operating systems, eliminating extraneous services in hosts and servers, eliminating default passwords and accounts, and configuring devices correctly can be accomplished through old-fashioned perseverance. Other vulnerabilities, caused by programming shortcomings, can be controlled by radical new procedures, such as independent code reviews. Still, managing other vulnerabilities created by remote-access scenarios may require a financial outlay for strong authentication, but in open-access environments, such acquisitions are potentially justified. The security policy should be penned after an exhaustive review with all strategic business concerns, because the policy should relate directly to current needs and projected requirements during the life of the policy.

The steps you should take to implement the suggestions in this chapter could possibly be implemented with available staff. If staff is limited, you might consider using scanning or intrusion detection tools to concentrate your focus. Regardless of your particular circumstances, falling short of accomplishing the suggestions given in this chapter will compromise the effectiveness of your e-security program before it is out of the blocks. Therefore, arm yourself with these countermeasures; you are in a new arms race, for certain.

PART III

Waging War for Control of Cyberspace

Up to this point, you have acquired a general understanding of what e-security entails and what's at stake: e-business. You also know that protecting your information assets is critical to achieving business goals in open-access environments. The extent to which you pursue e-business initiatives depends on how well you protect information assets and related networking infrastructure. In other words, the ultimate goal of your e-security program should be to enable e-business: alternative revenue streams, virtual supply chains, preemptive marketing advantage, strategic partnerships, and competitive advantage.

We now move from concept to reality and look at the developments, knowledge base, solutions, and status of e-security. Part III begins by reviewing the attack tools and weapons that hackers use to exploit common and not-so-common vulnerabilities (Chapters 6 and 7). Chapters 8–11 focus on surviving an attack: what to do in its aftermath and countermeasures needed for thwarting a cunning adversary and potential infiltrator (Chapter 8). Chapter 9 looks at dealing with distributed denial-of-service (DDoS) attacks. Chapter 10 discusses various countermeasures that can be deployed at various network-operating layers, such as the router and operating system. Chapter 11 discusses the deployment of security architecture in layers.

CHAPTER SIX

Attacks by Syntax: Hacker and Cracker Tools

This chapter reviews the tools, techniques, and strategies of hackers and crackers in detail. Hacker tools or weapons are easily accessible and typically obtained by downloading them from the Internet. In some cases, the hackers are well organized and managed, such as the Cult of the Dead Cow, authors of the Back Orifice series of backdoor programs. Organized hackers are the perpetrators we should be *very* concerned with. In other cases, they aren't so organized; instead, their attacks are random and opportunistic. Nonetheless, the effects of their attacks may range from mayhem, which causes lost worker hours to fight the spread of viruses or to restore defaced Web pages, to malicious, which includes DDoS and outright destruction of information assets and intellectual capital.

Inherent Shortcomings of TCP/IP

Much of the security woes of e-business channels can be attributed to the inherent shortcomings of TCP/IP, the underlying Internet protocol. The TCP/IP designers worked in low-security academic research environments. TCP/IP was developed for use with the ARPANET, the predecessor of the Internet. When TCP/IP was developed, security was not a major concern. The designers were interested mostly in developing an operating system that would be compatible across heterogeneous platforms but they were also charged with creating computing environments in which information could be freely shared without unnecessary restrictions. (Information wants to be free: Does this sound familiar?

> The Internet protocol suite, known as TCP/IP, was designed in low-security academic research environments in California and Massachusetts. In the early days (the Sixties), university computer departments provided a congenial environment where creativity flourished; openness and consideration for others were considered the norm. In this environment, some users considered security restrictions undesirable, because they reduced accessibility to freely shared data—the *hallmark of the community in those days.* Security restrictions make it more difficult to access data. What is the point of such restrictions, if access is inherently valuable?[1]

Therefore, there was a tradeoff of greater security in favor of greater support for heterogeneous systems.

Ultimately, only a base level of security was implemented in TCP/IP. For example, the protocol incorporates only user IDs and passwords to provide a rudimentary level of authentication. Also, *IP address screening* was built into the protocol to prevent users from accessing a network unless they come from trusted domains. Unfortunately, both measures were and still are ineffective. Passwords in TCP/IP systems are often easily guessed or intercepted with packet sniffers. IP address screening does little or nothing to thwart IP spoofing, primarily because this feature doesn't contain any mechanism for verifying the authentication of incoming data packets or ensuring that the packets are coming from the domains they *should* be coming from. In other words, if the IP address of a trusted domain is confiscated and is used to fake or to spoof the source of an IP transmission, TCP/IP has no way of verifying whether the data is coming from the trusted domain or elsewhere.

TCP/IP also yields security holes when it is improperly configured. For this reason, it is important that TCP/IP be configured to deliver only the services that are required by the applications of the network. Extraneous services, default settings, and passwords should all be eliminated before IP-based networks support production environments.

Unfortunately, with all its robustness, TCP/IP is a fundamentally insecure network architecture. Since inception, potentially hundreds of security holes are thought to be prevalent. Although many of them have been discovered and

---

1. Excerpted from *Building a Strategic Extranet*, Bryan Pfaffenberger, IDG Books, 1998, p. 91.

patched, others remain to be discovered and exploited. (For example, researchers at the University of Finland at Oulu discovered that through SNMP, a TCP/IP service for remote access and control, ISPs' network devices could be shut down or fully controlled by an attacker, depending on the flavor of SNMP. Apparently, this vulnerability has existed for more than 10 years.)

Generally, attacks can occur when data is en route or residing in host computers. With TCP/IP, any computer with a *legal* IP address is a host computer. Thus, from the perspective of hackers and especially disgruntled or former employees, TCP/IP provides potentially numerous illegal points of entry that can no longer be controlled in IP networks without the aid of *add-on* security measures. The next several sections look at specific classes of attacks that exploit TCP/IP weaknesses and flaws.

Standard "Ports" of Call

TCP/IP-based networks have approximately 130,000 IP ports, or doorways, for interactive communications among network devices, services, applications, and discrete tasks. IP ports have predefined purposes for protocols or services. Some ports used for common types of transmissions include:

Port	Protocol/Services
Port 21	FTP
Port 23	TELNET
Port 25	SMTP
Port 53	DNS
Port 80	HTTP (typical Web traffic)
Port 111	RpcBind
Port 113	AUTHd
Ports 137–139	Netbios
Port 1524	Oracle DBMS TCP/IP communications
Port 443	HTTPs

Ports that are not used regularly or as common are port 110: POP3 (Post Office Protocol version 3), port 389: LDAP (Lightweight Directory Access Protocol), and port 8080: Web site testing. Effective security translates to protecting all 130,000 doors within your network.

As you know, a security measure, such as a firewall, will come out of the box with all ports disabled. Consequently, the specific group of ports that your network applications require would be a matter of enabling those ports across the firewall. For a given network, this could mean enabling hundreds or even thousands of doors. For a firewall, however, the *number* of ports initialized for access doesn't pose any particular problem inasmuch as ports being available to begin with. As long as the doors, or ports, are opened, depending on the service or protocol, the network could be at risk to attacks that exploit vulnerable services or protocols that are accessed through the firewall. Therefore, proper diligence should be performed to eliminate as much vulnerability as possible by applying the latest security patches, keeping firewalls configured correctly, and deactivating them when no longer required by applications.

Because business needs continually change, the firewall rule base must also change to reflect the enterprise's current need. The firewall rule base is a working rule set that should reflect the current security policy of the organization. Inevitably, ad hoc business needs demand that certain other ports be enabled that may occasionally not be covered by an organization's current security policy. Under these circumstances, the proper diligence may not be performed, perhaps because the need for the service is temporary. Accommodating such requests, or when the appropriate security checks aren't performed, is often referred to as poking holes through the firewall. When holes are being poked through the firewall consistently, potential security risks may arise. To complicate matters, if vulnerabilities exist on these services, a skilled hacker or one with good tools can easily gain access through the best firewalls.

Port scanning is the hacker's favorite technique for gaining illegal access to networks. Hackers count on finding unwatched doors and windows through the firewall owing to misconfigured rules or vulnerabilities in the network. Firewalls show port-scanning activity as a series of connection attempts that have been dropped. In a port scan, the source address and the destination address—typically, the firewall—stay the same, but the destination IP port number changes in sequence because port-scanning tools/applications attempt connection through ports in sequence. Other port-scanning tools try ports randomly, to make detection more difficult.

Port-scanning tools are also fairly sophisticated. When a port appears to be unprotected, the tool logs the information for the hacker to investigate later. Hackers scan literally hundreds, if not thousands, of ports before coming across a poorly secured door. For example, say that the firewall supports access to ports 386, 387, 388, and 389 and that port 389 presents a vulnerable service that can be exploited. Most hackers would not attempt more than three or four random scans at a time, because they know that too many dropped sequential-access attempts against IP ports in the firewall logs would signify port-scanning activity. Hackers have patience and resolve, because finding an open door could take literally an indefinite amount of time.

TCP/IP Implementation Weaknesses

When IP networks transmit data over a wide area from a source to a destination host, three things occur.

1. The data traverses numerous routers, also called hops.
2. The IP packets are typically divided into smaller units, or IP fragments.
3. Depending on network traffic, a given IP fragment can travel a path different from that of another fragment.

Each of the smaller IP packets, or chunks, is a replica of the original IP packet, except that the chunk contains an *offset field,* created when TCP/IP deems it necessary to break down the original IP packet into smaller units. When the IP fragment arrives at the destination host, the offset field tells the host the number of bytes the field contains and their order of position in the original IP packet. This information enables the destination host to reconstruct the IP fragments into the original IP packet, or into the packet the *source* host transmitted.

For example, suppose that an IP packet contains 400 bytes of information and is transmitted as three IP fragments. The offset field of one arriving IP fragment tells the destination host that the field contains bytes "1 through 200." The offset fields of subsequent IP fragments would therefore contain bytes "201 through 300" and "301 through 400." Because IP fragments won't necessarily arrive in order, the offset field ensures that IP fragments are reconstructed into the proper sequence. Traveling in smaller units also increases network throughput and minimizes the effects of latency in routers, switches, and other network devices.

Ping of Death

The ability to reduce IP packets into smaller units is a nice feature of TCP/IP communications. Unfortunately, this native feature is also an inherent weakness. Hackers have devised several techniques to exploit this problem. Using ping of death and such variants as Teardrop, Bonk, and Nestea, hackers disrupt the offset field's ability to align IP fragments properly during the reassembly process (see Figure 6–1).

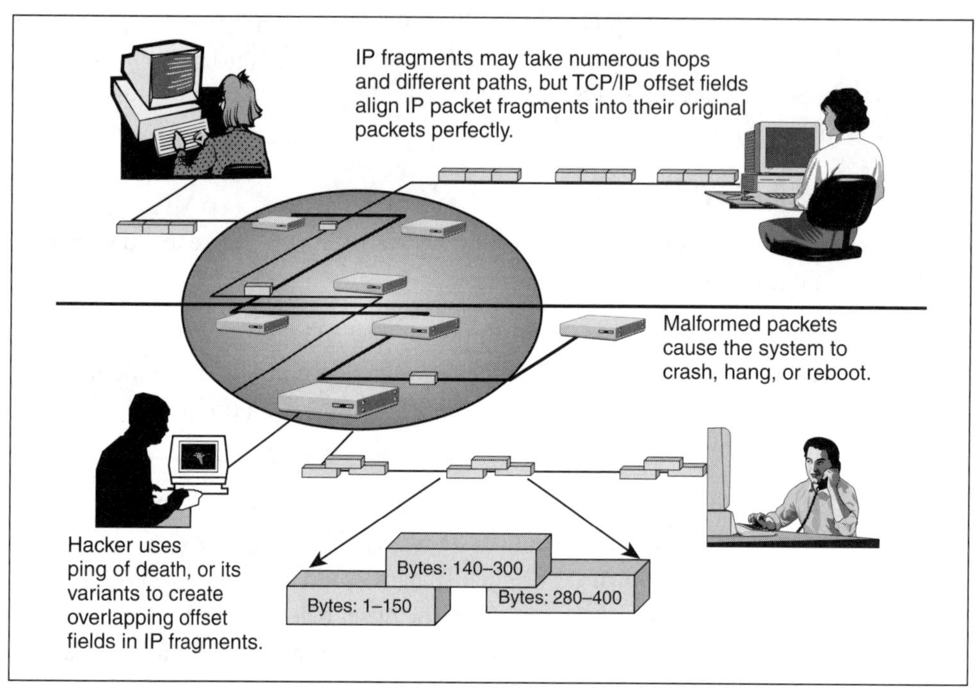

Figure 6–1 Exploiting TCP/IP weaknesses with ping of death

These tools enable hackers to reduce IP packets into fragments with *overlapping* offset fields. Therefore, when they are reassembled at the destination host, the overlapping offset fields force IP to reconstruct them into malformed IP packets. The malformed packets don't create any serious damage but could cause hosts to crash, hang, or reboot, resulting in lost data and time. Although these security issues are not serious, they are hassles you could live without. Most firewall-based security measures block ping-of-death incursions. Make certain that they can also handle the other variants.

SYN Flood and Land Attacks

One of the hacker's favorite tactics is using the initial session establishment between client/server applications to launch an attack. Communications between an initiating and receiving application— a TCP session—occurs as follows.

1. To begin a TCP session, the initiating application transmits a *synchronize* packet (SYN) to the receiving application (host).

2. The receiving host responds by transmitting a synchronized acknowledgment packet (SYN-ACK) back to the initiating host.
3. To complete the connection, the initiating host also responds with an acknowledgment (ACK). After the handshake, the applications are set to send and to receive data. (See Appendix D for a detailed description of SYN-ACK attack.)

Metaphorically, if SYN flood is attack by "sea," or flooding, the land attack is just the opposite. In this scenario, the destination address of the receiving host is also the *source* IP address. In other words, the SYN packet's source address and target address are the same. So when it tries to respond by sending a SYN-ACK, the receiving host tries to respond to itself, which it can't do. Land attacks are another denial-of-service ploy; the targeted application will ignore all legitimate requests while futile attempts to respond to itself are continued.

Firewalls handle SYN floods by sending the final ACK and monitoring the connection to determine whether normal communications are conducted. If nothing transpires, the connection is terminated. Land attacks are prevented by anti–IP spoofing measures. Typically, when trusted internal IP addresses originate on external ports, antispoofing features will automatically drop the connections, thereby thwarting any land attacks.

IP Spoofing

Practically all IP packet–based attacks use IP spoofing, especially basic denial-of-service attacks, such as SYN-ACK, ping of death, and land attacks, and the more sophisticated distributed denial-of-service (DDoS) attacks, which depend on master/slave relationships to function properly. IP spoofing is popular because it hides the hacker's identity and provides the means to slip into your network. IP spoofing works only when your network or security measures believe that the source address of the IP packet originates from a trusted domain. (See the discussion in Chapter 4.) The main method of thwarting IP spoofing is to use security measures that reject packets when *trusted* or *internal* IP source addresses arrive on *external* ports.

IP spoofing combined with DDoS incursions creates a formidable attack. Although properly configured firewalls can recognize all "flooding" or variations of DoS attacks, once the firewall has been breached, the hacker gains a foothold into your network. When this happens, the firewall won't help you identify the source of the attack, because the address is faked. However, the firewall will log the suspicious traffic. With this information, you can work with your ISP to help filter out the bogus traffic before it does much harm to your network. (Attack prevention and survival are discussed further in Chapter 8.)

Distributed Denial-of-Service Attacks and Tools

Everyone was concerned that the new millennium might usher in the Y2K bug. Many braced for the apocalypse, but what we got amounted to no more than a *cloudy day with scattered showers.* As the clouds passed and the world was just beginning to breathe a sigh of relief, several well-known business entities were struck by the computer world's equivalent of a flash flood: a distributed denial-of-service (DDoS) attack. Amazon.com, Yahoo, and E-Trade led the distinguished list of companies that were brazenly attacked one morning in early February 2000.

A DDoS attack is a coordinated, militaristic attack from many sources against one or more targets. In fact, that is the real major difference between a *distributed* DoS and a *regular* DoS. Other than the attack coming from many sources (distributed), both variations flood your network with such a high volume of useless packets that legitimate users can't get through. Usually, the hackers download DDoS tools from the Internet. Examples of such attack tools are Trin00, Tribe Flood Network (TFN), and TFN's latest version: Tribe Flood Network 2000 (TFN2K).

For most businesses, DDoS attacks would cause an inconvenience or loss of productivity. However, for e-business concerns, DDoS attacks could create substantial losses resulting from lost sales and customer confidence. Just ask E-Trade. (See the section Real-World Examples in Chapter 1.) The most important fact about DDoS attacks is that they are not designed to penetrate, destroy, or modify your network but to bring you down, perhaps indefinitely. This is not sporting at all, as if any hacker exploit could be.

DDoS attack tools are a hacker's dream, enabling intruders to attack with the element of surprise by mobilizing forces covertly and with precision. DDoS attacks exploit arguably the single greatest advantage afforded by the Internet: distributed client/server functionality. In general, DDoS attack tools could not work without this important feature, because they depend on client/server relationships between the hacker and the *master* and between the master and the attack tools themselves, the *daemons.*

After finding a way into your network, usually by exploiting vulnerabilities or unwatched doors, the hacker installs daemons on the compromised host.

Daemons are simply software utilities that service the requests of the master program by initializing the hosts to send large and useless packets of information to an unsuspecting target. Once the host is compromised, the host becomes an agent, and the daemon becomes a zombie, which is *dead* until summoned *to life*—into action—by a master. Once hundreds or perhaps thousands of zombies are in place, the DDoS attack is orchestrated. The hacker programs the master to command the zombies to launch a DoS attack against a single unsuspecting target: simultaneously. A single DoS attack would not do much to a network with high-bandwidth Internet access, but thousands of these attacks originating from around the globe would effectively overwhelm a site and deny service to legitimate users.

Table 6–1 summarizes the salient features of DDoS attack tools. (Also, see Appendix D for a detailed description of DDoS attack tools.)

Trin00

One of the most powerful DDoS attack packages is Trin00, or Trinoo. Trinoo made its debut toward the end of 1999 by attacking several high-capacity commercial and university networks. Unlike the other DDoS attack tools, Trinoo does *not* spoof the source addresses during the attack. The source addresses of the attacks are the hosts that were compromised by Trinoo daemons. The hosts can belong to anyone anywhere on the Internet. The type of flood that Trinoo dispenses is a UDP flood. The DoS occurs when the target host becomes inundated with a flood of UDP packets to process while denying service to legitimate operations.

The orchestration of Trinoo is more a test of intestinal fortitude than of skill. Through diligence and patience, the intruder scans literally hundreds, perhaps thousands, of Internet hosts for vulnerabilities before *a legion* of hosts are commandeered. Setting up this attack is also a testament to client/server functionality and distributed processing. Using vulnerable hosts, the master and daemon programs are clandestinely installed one after another until all Trinoo's components are in position. Several of the hosts are used for the "master programs," and each master controls a cluster of hosts that have been invaded by Trinoo's daemons. Figure 6–2 summarizes Trinoo.

Tribe Flood Network

Like its cousin Trinoo, TFN is an attack tool that is able to launch DDoS attacks from multiple locations against one or more targets. Unlike Trinoo, the source IP addresses of the attack packets can be spoofed. Hackers like TFN because it can generate a variety of DoS attacks. In addition to a UDP flood, TFN is capable of

Table 6–1 Comparison of DDoS Attack Tools

	Trin00	TFN	TFN2K	Stacheldraht
Client/Server Architecture	3 tier: client, master, daemons	2 tier: client, daemons	2 tier: client, daemons	3 tier: client, handler, agent
DoS Flood Type (Attack)	UDP flood	UDP, ICMP echo request, SYN-ACK, Smurf bandwidth	UDP, ICMP echo request, SYN-ACK, Smurf bandwidth	UDP, ICMP echo request, SYN-ACK, Smurf bandwidth
Operating System	UNIX and NT	UNIX and NT	Linux, Solaris, and NT	Linux and Solaris
Intruder Port (to Master)	TCP: TELNET, NETCAT: interface is from a client	N/A (Interface directly into master)	N/A (Interface directly into master)	TCP: TELNET Alike (Interface is from a client)
Master Port (to Daemon)	UDP	TCP, UDP	TCP, UDP, ICMP	TCP, ICMP
Daemon Port (to Master)	UDP	TCP, UDP	TCP, UDP, ICMP	TCP, ICMP
IP Spoofing	No	Yes	Yes	Yes
Encryption	List of known daemons in file master	IP addresses in daemons with Blowfish	Data fields are encrypted, using the CAST algorithm	• TELNET Alike Session • Master and Daemon links with Blowfish

generating a TCP SYN flood ICMP echo request flood, and an ICMP directed broadcast, also known as a Smurf bandwidth attack. Based on a client/server relationship, the TFN intruder interfaces to a TFN master with a character-based command line to provide attack instructions to a legion of TFN daemons.

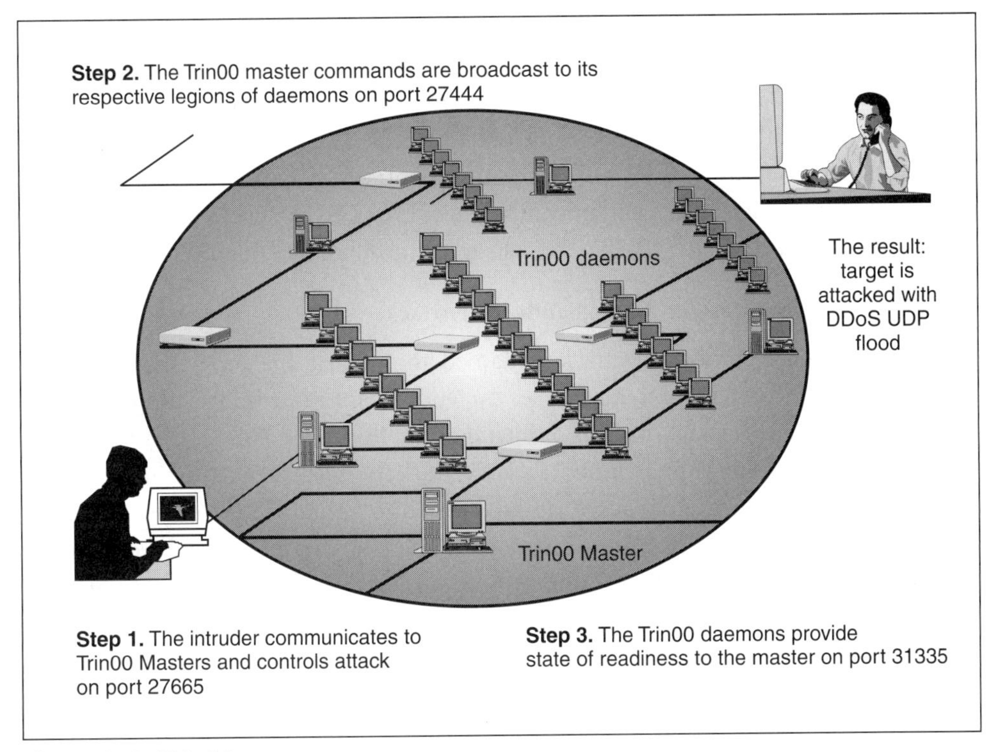

Figure 6–2 Trin00

Tribe Flood Network 2000

TFN2K, a descendant of TFN, uses client/server architecture. Like TFN, TFN2K operates under UNIX-based systems, incorporates IP address spoofing, and is capable of launching coordinated DoS attacks from many sources against one or more unsuspecting targets simultaneously. TFN2K can also generate all the attacks of TFN.

Unlike TFN, TFN2K also works under Windows NT, and traffic is more difficult to discern and to filter because *encryption* is used to scramble the communications between master and daemon. TFN2K is also more difficult to detect than its predecessor because the attack tool can manufacture "decoy packets" to non-targeted hosts.

Unique to TFN2K is its ability to launch a Teardrop attack. Teardrop takes advantage of improperly configured TCP/IP IP fragmentation reassembly code. In this situation, this function does not properly handle overlapping IP fragments. When Teardrop is encountered, systems simply crash.

Stacheldraht

German for barbed wire, Stacheldraht debuted on the Internet between late 1999 and early 2000, compromising hundreds of systems in both the United States and Europe. Although this DDoS tool looks a lot like TFN, its architecture resembles Trinoo's, which is based on a three-tier client/server approach, with client (intruder), master, and daemon components. Like TFN, Stacheldraht can generate UDP, SYN-ACK, ICMP echo request, and Smurf bandwidth DDoS attacks.

ICMP Directed Broadcast, or Smurf Bandwidth Attack

The Smurf bandwidth attack is named for one of the programs in the set required to execute this attack. (See Appendix D for more information on Smurf attack.) In this attack, hackers or intruders use the inherent characteristics of two services in TCP/IP to orchestrate a DoS attack for unsuspecting victims.

In IP networks, a packet can be directed to an individual host or *broadcast* to an entire network. Direct broadcast addressing can be accessed from within a network or from external source locations. When a packet from a local host is sent to its network's IP broadcast address, the packet is broadcast, or delivered to all hosts on that network. Similarly, when a packet that originates from outside a local network is transmitted to a network's IP broadcast, the packet is also sent to every machine on the local or target network. In general, make sure that your network filters out ICMP traffic directed to your network's broadcast address. (See Chapter 9 for more on prevention of Smurf bandwidth attacks.)

Backdoor Programs and Trojan Horses

Technically speaking, a backdoor program is not a Trojan horse, although the two terms tend to be used synonymously. A Trojan horse is a malicious program that is concealed within a relatively benign program. When a victim runs the innocuous program, the Trojan horse executes surreptitiously, without the victim's knowledge. By contrast, a backdoor program sets forth the notion of circumventing a regular entry or access point, such as through a user "front door," or login routine. A back door, an entry point into a host computer system, bypasses the normal user login procedures and potential security measures of that system. A back door could be a hardware or software mechanism but is generally software based; like a Trojan horse, it is activated secretly, without the victim's knowledge.

An intruder's favorite method of slipping backdoor programs into trustworthy hosts is to send the backdoor program as an e-mail attachment with an innocent-sounding name. Because of this trick, a backdoor program is referred to as a Trojan horse. The common element in both a backdoor program and a Trojan horse is that they function without the user's knowledge.

The ultimate goal of the more than 120 backdoor programs existing in the wild is to secretly control your computer system from a remote hiding place. In order for a backdoor program to function properly, it must gain root access to the host system. Root access allows an intruder to do what he or she wants in your computing environment. Once a backdoor program is in place, your system can be accessed and manipulated, unbeknownst to you, whenever the intruder desires. That's the unfortunate reality of backdoor programs.

Backdoor Program Functions

Remote-controlled backdoor programs can be instructed to perform a litany of functions, ranging from cute annoyances to outright maliciousness. To annoy you, backdoor programs can open and close your CD-ROM drive, reboot your machine, start your screensaver automatically, reassign your mouse buttons, shut down the computer, play sounds, and even look through a Webcam on your system. However, the main purpose of backdoor programs is for malicious intent and misuse. They can steal your passwords by monitoring fields in dialog boxes or keystrokes, regardless of whether they are encrypted, plain text, dial-up, or cached. Backdoor programs destroy and modify data and files, deleting system files, moving and killing windows on your desktop, hijacking communication sessions by port redirection, executing files, reading and writing to the registry, and capturing screenshots.

One of the more insidious features of these programs is their ability to use your computer as a server, such as an FTP server, to support illicit activities. Finally, backdoor programs can open your Web browser to any URL, send messages by pretending to be you, and play proxy for you in chat rooms through Internet Relay Chat (IRC) connections. The ramifications of the last exploit can be significant. An intruder, masked by your Internet account, can engage in undesirable activities, such as use of vulgar or illicit language, that, at a minimum, can prompt a reprimand from the chat room "police" or make you liable for harassment.

Examples of Backdoor Programs

The most popular backdoor programs in use in the wild are Back Orifice 2000 and Netbus 2.0 Pro (NB2). Back Orifice 2000 is the newest release of Back Orifice, which made its debut in July 1998 as the brainchild of the Cult of the Dead Cow, a hacker organization that has vowed to thwart Microsoft at every opportunity. Netbus has been widely accepted and in use for some time. Both Back Orifice 2000 and Netbus are compatible with Windows 95, 98 and NT operating systems. Table 6–2 summarizes these and other popular backdoor programs.

Like other backdoor programs, Back Orifice 2000 gathers information, performs system commands, reconfigures machines, and redirects network traffic. However, Back Orifice 2000 appears to be the granddaddy of them all, with more

Table 6–2 General Characteristics of Some Popular Backdoor Programs

	File Name				
Program	Preinstall	Postinstall	Attacker Link	Listening Ports	Encryption Use
Back Orifice 2000	• bo2K.exe (server) • bo2kgui.exe	• FileUMGR32.exe • Remote administration service (NT)	TCP/UDP	• UDP 31337 • Optional	Client/ Server link: 3DES
NB2	N/A	NbSvr.exe	TCP	20034	Client/ server link: weak
DeepThroat	N/A	Systemtray.exe	UDP	• 2140 • 3150	N/A
NetSphere	N/A	Nssx.exe	TCP	• 30100 • 30102	N/A
GateCrasher	N/A	System.exe	TCP	• 6969	N/A
Portal of Doom	N/A	Ljsgz.exe	UDP	• 10067 • 10167	N/A
GirlFriend	N/A	Windll.exe	TCP	• 21554	N/A
Hack'a'Tack	N/A	Expl32.exe	TCP/UDP	• 31785 TCP • 31789 UDP • 31791 UDP	N/A
EvilFTP	N/A	Msrun.exe	TCP/FTP	• 23456	N/A
Phase Zero	N/A	Msgsvr32.exe	TCP	• 555	N/A
SubSeven	N/A	Explorer.exe	TCP	• 1243 • 6711 • 6776	N/A

than 70 commands at an intruder's disposal. Its most compelling feature is that once installed, it becomes practically invisible.

Back Orifice 2000 has been used as a simple monitoring tool, but its primary use appears to be to maintain unauthorized control over a target host to commandeer resources, render vexing annoyances, steal passwords, and generally collect and modify data. This program also has an array of commands to eavesdrop, play, steal, and manipulate multimedia files. Moreover, it supports adjunct programs called *plug-ins,* to enhance its overall capability.

Netbus 2.0 (NB2) descended from the popular backdoor program Netbus. NB2 does pretty much everything Back Orifice 2000 is able to do: manipulate files, provide full control over all windows, capture video from a video input device, and support plug-ins. NB2 also enables the attacker to find cached passwords and to run scripts on a specified host(s) at a given time. NB2 supports an "invisible mode" to hide from users of infected machines. It also possesses a peculiar feature whereby it will notify users of compromised machines after installation. However, attackers can hide this feature with relatively little modification.

Summary

The most important fact about backdoor program incursions is that once they are installed on your system, the original attacker(s) can make the back door available to others. The implications are frightening. Another important point to remember is that backdoor incursions are not dependent on exploiting any inherent or related vulnerability in operating systems or applications. Usually, the backdoor program finds its way into a machine by trickery or through the use of a Trojan horse. In other words, it finds its way into your system as an e-mail attachment, for example, with an innocent-sounding name.

Fortunately, backdoor programs are easily disposed of once they are detected, but because of their chameleonlike characteristics, detection is difficult. To determine whether a back door is active on your system is a straightforward process if you know what to do. If the back door is controlled by a UDP connection, it requires sending a UDP packet to that port, with the appropriate specifications, and the backdoor program in question will respond. If the backdoor program communicates by TCP, you can TELNET to the suspicious port and provide the necessary information, which should also provide you with a response from the backdoor program in question.

To eliminate backdoor programs manually, you must obtain the appropriate instructions. Therefore, you should make it a regular practice to stay in touch with several security organizations that disseminate alerts and up-to-the-minute information on coping with security issues and threats. Carnegie-Mellon's CERT Coordination Center (CERT/CC), SANS Institute Resources, Internet Security Systems' "X-Force" alerts, Axent Technologies' "SWAT," and Web repositories, such as NT BugTraq Web site, are all good sources of finding step-by-step instructions for eliminating or mitigating the effects of security threats and risks.

If an attacker renamed the backdoor program, the step-by-step instructions may not be sufficient to rid yourself of a particular problem. In this case, you have to rely on a host and network vulnerability assessment and intrusion detection system and a strong virus protection implementation.

CHAPTER SEVEN

Attacks by Automated Command Sequences

Viruses are typically spread through e-mail with infected attachments or media, such as floppy disks. Distributed denial-of-service attacks take advantage of inherent weaknesses in TCP/IP and related vulnerabilities in network devices, such as firewalls, hosts, and routers. Backdoor programs typically find their way into a network as a Trojan horse or an innocuously appearing component of a larger, perhaps more recognizable, program. The one common thread in these three classes of incursions is that they have been active in the wild for sometime. These incursions also have identities of their own, with distinct signatures, or command structure, or syntax.

In this chapter, we explore another mode of intrusion: script attacks. Unlike DDoS and backdoor incursions, which must be orchestrated by an intruder, a script attack directs itself through automated command sequences. The code of a script that perpetuates a specific attack is continually being modified to produce other attacks. When unleashed, script attacks steal passwords, credit card numbers, and sensitive information; modify and destroy data; hijack sessions; and, in some cases, enable alternative pathways into a trusted network domain. In this regard, a script attack functions like your *garden-variety* backdoor program. In other circumstances, script attacks function like a virus, when data is destroyed, modified, or changed or when applications perform erratically. This chapter also explores other sources of vulnerabilities that, because of their pervasiveness, your life-cycle security measure(s) should address when instituted within your enterprise.

Script Attacks

A script is a small, self-contained program that performs specified tasks within client/server applications. It gives developers the flexibility to build functionality into applications whenever they desire. Scripts usually provide client-side functionality; when a user views a Web site, for example, scripts are downloaded with the page and begin executing immediately. This is a key reason that hackers like scripts. In contrast, when plug-ins are downloaded, the user must take the time to install it and perhaps restart the computer.

Scripts are developed with script languages, such as JavaScript or Visual Basic Script (VBS), created by Netscape Communications and Microsoft, respectively. In general, JavaScript, which is a derivative of Java, allows Web designers to embed simple programming instructions within the HTML text of their Web pages. Both Microsoft's Internet Explorer and Netscape Communicator support JavaScript. On the other hand, VBS, a scripting language based on Visual Basic, provides the macro language used by most Microsoft applications. Like JavaScript, VBS scripting is also used in Web page development. However, it is compatible with Netscape communicator with a plug-in.

Using script languages to create attacks or malicious code is quite popular among hackers because the potential for attack variations is virtually unlimited. Script attacks typically are sent via e-mail. But other malicious code can be encountered through rogue Web sites, another name for Web locations with embedded malicious code. A script is easily modified, so any given one can perform a variety of attacks. More important, each time malicious code is modified, its signature or command structure, or syntax, changes, making *individual* script attacks, and particularly their derivations, difficult to deal with through current security measures. When the Love Bug was released onto approximately 45 million computers in the wild, a deadlier but slower-moving variation, called New Love by the media and Herbie by the Justice Department, materialized within days of the Love Bug's demise. Within 2 weeks of Herbie's debut, about 29 variations of the Love Bug were spawned.

In other words, when a script attack is launched, newer versions of the attack are typically unleashed into the Internet community faster than the white hats can release countermeasures. The main problem with security measures is that they usually can't protect against *new* script-based viruses until the virus has been fingerprinted and placed in their signature database—in the case of an intrusion detection system—or data files—in the case of a virus protection system. However, no sooner than a culprit has been corralled and fingerprinted, thwarting subsequent incursions is relatively easy to do. Recognizing the problem, companies that supply security measures are continually on the lookout for new threats in the wild, including e-mail-borne script viruses. As soon as they are discovered and fingerprinted, they can be added as updates to the respective client bases in a timely fashion. This keeps clients a little ahead of the game.

Another alternative to address-embedded scripted attacks is to harden network applications by disabling automatic scripting capabilities. Some products activate powerful scripting capabilities by default. This feature is called *active scripting* in Microsoft applications. Windows Scripting Host (WSH) in Windows 98 provides the functionality.

The majority of users do not require or want embedded scripting enabled, especially WSH in Microsoft applications. For example, WSH is installed and activated by default with Windows 98 and Internet Explorer version 4.0 and higher. With WSH enabled, users can execute .VBS files by double clicking. In contrast, a new generation of scripted e-mail attacks will execute *without* user intervention, through viewing only. Therefore, unless certain users need to execute embedded scripting automatically, the best bet is to disable scripting capabilities. This will disable both VBScript and JavaScript and default functions, such as Windows Scripting Host in Windows operating environments.

The Next Generation of E-Mail Attacks

The next generation of scripted e-mail attacks does not require the user to be duped into double clicking the script—malicious code—for execution. For example, if you are using Outlook Express with the Preview Pane enabled, this scripted e-mail attack can infect a host without the user's ever opening the e-mail. This class of attack is the next generation of e-mail *worms.*

A worm is a virus that propagates itself without the aid of another program or user intervention. Worms are also nefarious. The Love Bug's offspring, Herbie, was able to change the subject line of the e-mail whenever it propagated itself from host to host as an e-mail attachment. This polymorphous nature of worms makes them very difficult to detect with content scanners or virus software. Although worms, such as Herbie and the Love Bug, are self-replicating and destructive at the same time, the user still must execute the embedded script in the e-mail to activate the worm.

In contrast, next-generation scripted e-mail attacks will not require the user to double click embedded script in the e-mail for it to propagate. The attack merely needs to arrive and take advantage of certain vulnerabilities in the target host. As mentioned, if Outlook Express has the Preview Pane activated, coupled with embedded scripting, the next-generation worm can literally wreak havoc. It arrives by stealth, becomes activated automatically, exacts damage in a variety of ways, disguises itself, secretly targets everyone in the user's e-mail address book, and reproduces the attack over and over again at every receiving host. Scary indeed!

The Bubble Boy Virus

The Bubble Boy virus was the proof of concept of the dangerous next-generation self-activating, self-propagating scripted e-mail attacks. If active scripting is enabled and the Preview Pane is activated in Outlook Express, for example, simply viewing the infected e-mail will launch the virus attack. In other words, the user does not have to physically open the embedded script—attachment—in the e-mail to activate the attack.

The Bubble Boy virus borrows its name from an episode of the popular *Jerry Seinfeld* show. In that episode, Jerry's sidekick, George, plays *Trivial Pursuit* with a boy confined to an oxygen bubble because of a faulty immune system. The boy rightfully answers "Moors" to one of the questions, but the game card misspelled it as "Moops." When George insists that the answer is Moops, a heated argument unfolds, and a fight ensues. To everyone's dismay, the fight ends when George accidentally pops the oxygen bubble.

The Bubble Boy virus quietly made its debut on the Internet in late 1999, attacking Microsoft's Outlook e-mail system. Classified as a worm, the virus sent itself to everyone listed in the infected system's e-mail address book. The Bubble Boy virus spread itself to everyone listed in the victim's address book. This was the extent of the attack. However, the attack could have just as easily deleted, modified, or stolen data before it moved on. More significantly, the attack proved that a scripted e-mail attack could occur by stealth, without any user action(s) whatsoever.

The implications of this development bode potentially disastrous consequences for the Internet community. One dangerous possibility involves a self-replicating distributed denial-of-service attack tool that spreads through e-mail. Imagine being bombarded by a DDoS attack from every recipient of an e-mail worm running Outlook or Outlook Express. Even worse, imagine being bombarded with a DDoS attack from the 45 million recipients of the Love Bug virus worldwide.

Mainstream JavaScript Attacks

Although most of the discussion in this chapter has focused on script attacks involving Microsoft's VBS language, comparable JavaScript-based attacks have also emerged into mainstream scrutiny. The irony of it all is that certain of these JavaScript attacks exploit vulnerabilities in Microsoft applications.

JavaScript Attacks on Microsoft's E-Mail Programs

If your enterprise uses Outlook 98, Outlook Express 5, or Outlook 2000, you should be aware of an embedded e-mail JavaScript attack on the prowl in the wild. Although Microsoft has developed and issued patches to correct the vulnerability, these e-mail applications, if improperly configured, are still susceptible to the threat, which directs them to perform unauthorized activity.

The attack begins after the e-mail program receives instructions from the malicious JavaScript code. The Outlook-based client opens a browser window to any URL of the intruder's choice. The unwitting clients tend to be compromised in two ways. Submitting data to complete forms on Web sites is one attack; the other one, which is more diabolical, involves directing the user to a Web site that can load Web pages capable of exploiting vulnerabilities that are not exploitable by e-mail attacks. This embedded JavaScript incursion can be used in conjunction with other exploits, such as one that levels attacks on cookies that are created while the user is surfing with Internet Explorer.

Internet Explorer Cookies Crumble

Cookies are one of the necessary evils of doing business on the Internet. A cookie, primarily a marketing tool, is an inherently intrusive practice. Web developers use cookies to collect information on Internet surfers when users visit or purchase anything from a Web site. Cookies enable Web merchants to ascertain what users do at a site, chart their patterns, and how often they visit. The betrayal comes when site managers combine this and other registration information into a demographic database of fellow surfers to sell to marketers.

Many surfers don't realize that a cookie is automatically created when they register at or merely visit a site.

When cookies are created, they reside on *your* system. Certain e-business Web sites, such as Amazon.com, Yahoo, NYTimes.com, and literally thousands of others, may even use cookies to authenticate users or to store confidential information. With the information provided by the cookie, the Web site registers you to facilitate interaction in subsequent visits. Although browsers allow you to disable cookies, some sites may function erratically or crash when they are visited. Others may require you to register again before browsing their sites, and you will lose the use of certain features, such as greetings with customized welcomes.

If cookies weren't irritating enough, Microsoft's Internet Explorer has given us something else to worry about: copycat exploits. Although Microsoft has issued a patch to fix the problem, be on the lookout for this exploit. Also, scripts can be used to take advantage of this vulnerability. In particular, derivations of a JavaScript-based attack could be prowling the Internet today.

Without the specific Microsoft patch that corrects this problem, all versions of Internet Explorer, including versions for Windows 95, 98, NT, and 2000 are vulnerable to this attack, as are versions for Solaris and HP-UX (Hewlett Packard-UNIX). Apparently, using a specially constructed uniform resource locator (URL), a rogue Web site can read any of the cookies in Internet Explorer clients that were created while browsing. For example, any cookies set from a domain that uses them for authentication or storing private information, such as Amazon.com, NYTimes.com, Yahoo Mail or MP3.com, can be read. Whenever any of the Internet Explorer clients encounters a specially constructed URL that references a cookie in the user's cookie file, IE gets confused and thinks that it is interacting with the legitimate URL and allows the information contained in the cookie in question to be read or modified. The description of this problem and patch can be found in security bulletin ms00–033.asp and obtained from Microsoft's Web site: http://www.microsoft.com/technet/security/bulletin/ms00–033.asp.

Netscape Communicator Cookie Hole

In Netscape Communicator version 4.x, a rogue Web site can set a cookie that allows a user's HTML files, including bookmark files, and browser cache files to be read. The exploit is accomplished with JavaScript code, which is included in the data that is composing the cookie. In order for this exploit to succeed, the Web surfer must have both JavaScript and cookies enabled, and Netscape Communicator's Default user profile must be active. To eliminate this problem, kill the Default user profile after setup and testing. This exploit depends largely on the intruder's being able to guess your Profile name in Netscape Communicator. The elimination of the Default user profile would not be enough if the intruder is able to guess your current user profile, which is normally a first name or user name portion of an e-mail address. To be safe, also disable JavaScript and/or cookies to preclude any chance of this exploit's working against you.

Attacks through Remote Procedure Call Services

Operating under TCP/IP, remote procedure calls (RPCs) are a useful and established network service that provides interactivity between hosts. RPC is a client/server utility that allows programs on one computer to execute programs on another computer. RPCs are widely used to access network services, such as shared directories, available through Network File System (NFS). Among other things, NFS allows a local user to map shared directories on remote hosts such that the directories appear as extended directories of a local host.

Over time, RPC services have gained a reputation for being insecure. Hackers discovered that although used primarily between remote hosts in an *internal* network, RPC enables *external* hosts to access internal networks by exploiting RPC vulnerabilities in internal hosts. For their first incursions, hackers used RPC to obtain password files and to change file permissions. In the much-publicized DDoS attacks that affected e-businesses in February 2000, evidence is compelling that the systems used in the attack were commandeered through vulnerabilities in RPC. For example, daemons belonging to both Trinoo and Tribe Flood Network DDoS attack tools are known to exploit these vulnerabilities, although these attack systems may not have necessarily been used in the brazen DDoS incursions.

Attacks through RPC vulnerabilities have become so exploited that they are third on the SANS Institute's list of the 20 most critical Internet security threats. (For a complete list, see Appendix A.) The following RPC services pose the most serious threats to networks:

- *Buffer overflow vulnerability in rpc.cmsd.* The Calendar Manager Service Daemon, or rpc.cmsd, is frequently distributed with the Common Desktop Environment (CDE) and Open Windows. This vulnerability enables remote and local users to execute arbitrary code or scripts with root privileges. The rpc.cmsd daemon usually operates with root privileges.
- *Vulnerability in rpc.statd exposes vulnerability in rpc.automountd.* Hackers use the vulnerabilities in these two programs together to attack internal hosts from remote Internet safe houses. The rpc.statd program communicates state changes among NFS clients and servers. The vulnerability in this RPC service allows an intruder to call arbitrary RPC services with the privileges of the rpc.statd process. In other words, the intruder can exploit any RPC service that may be called by the rpc.statd process. Typically, the called or compromised service may be a local service on the target machine or a network

service on another host within that same network. The rpc.automount program is used to mount certain types of file systems. This program allows a *local* intruder or internal saboteur to execute arbitrary commands with the privileges of the automountd process.

Both vulnerabilities have been prevalent for some time, and vendors that supply RPC services provide patches for these vulnerabilities. Too often, however, vendor-supplied patches are not applied. By exploiting these two vulnerabilities simultaneously, an interesting attack, to say the least, occurs. A *remote* intruder is able to relay, or bounce, RPC calls from the rpc.statd service to the automountd service on the same targeted machine. The rpc.automountd program does not normally accept commands from the network. Through this exploit, however, it accepts commands from not only the network but also sources that are external to the network. Once the connection is made, the intruder is able to execute arbitrary commands, including scripts, to attack the host, with all the privileges provided by automountd.

- *Vulnerability in ToolTalk RPC service.* The ToolTalk service allows independently developed applications to communicate with one another. ToolTalk's popularity hinges on the fact that application programs can interact through common ToolTalk messages, which provide a common protocol. Additionally, ToolTalk allows programs to be freely interchanged and new programs to be plugged into a system with minimal configuration.

  Hackers attack hosts that support ToolTalk services, exploiting a bug in the program's object database server. This vulnerability enables intruders to function as a super- or administrative user and to run arbitrary code or scripts on many mainstream UNIX operating systems supporting CDE and Open Windows.

Although various RPC services are being compromised, intruders are launching similar attacks. For example, any of the RPC vulnerabilities can be exploited to execute a malicious script with similar commands for inserting a privileged back door into a compromised host.

Certain forensics are associated with RPC exploitations. In general, you may discover the following kinds of activity:

- Core files for the rpc.ttdbserverd—for the ToolTalk database server—left in the root / directory in an attempt to attack rpc.ttdbserverd.
- Files named callog* discovered in the Calendar Manager Service Daemon spool directory, resulting from an attack on rpc.cmsd.

- Scripts that automate exploitations, which take advantage of privileged back doors. This method has been used to install and to launch various intruder tools and related archives, to execute attacks on other network hosts, and to install packet sniffers for illicit data gathering.

You may also encounter two archive files: neet.tar and leaf.tar. The neet.tar archive includes a packet sniffer, named update or update.hme, a backdoor program, named doc, and a replacement program, called ps, to mask intruder activity. In leaf.tar, you should find a replacement program, called infingerd, which creates a back door; an Internet Relay Chat (IRC) tool, called eggdrop; and related scripts and files. RPC vulnerabilities are also exploited to disseminate DDoS attack tools and to remove or to destroy binary and configuration files.

In any event, if you believe that a host has been compromised, it should be removed from the network immediately and steps taken to recover from a root compromise. Also, assume that user names and passwords have been confiscated from output logs and trust relationships with other hosts established. The Carnegie-Mellon CERT Coordination Center provides excellent guidelines for recovering from a root compromise. This document can be obtained at http://www.cert.org/tech_tips/root_compromise.html.

After recovering from the root compromise, RPC services should be turned off and/or removed from machines directly accessible from the Internet. If you must run these services, obtain patches from the vendor from which you purchased your host platform. Patches are available for IBM's AIX, Sun Microsystems' SunOS, SCO UNIX, HP-UX, Compaq's Digital UNIX, and so on.

Brown Orifice

Brown Orifice, a recently discovered attack that has surfaced in the wild, is dangerous because it attacks without the user's intervening, through trickery or other means. A user can be attacked simply by encountering a rogue Web site while surfing the Net. Surfers who browse with all versions of Netscape Navigator and Netscape Communicator version 4.74 and earlier are predisposed to attack. These versions include systems running Windows 2000, NT, and Linux. The problem is patched, however, in versions 4.74 and higher or version 6.0.

When encountered, Brown Orifice can initiate a series of commands that will allow a Java applet included in the browser to display a directory of what's on the surfer's hard drive. In other words, a surfer's standard PC can be tricked into thinking that it is a Web server capable of displaying the contents of its hard drive. This exploit allows the intruder to access any local file created by the user or any shared network files that are mapped to the user's machine.

Brown Orifice had attacked nearly 1,000 machines in practically no time at all. Reports of its attacks have been conflicting. One report says that the attacker can see, run, and delete files in the affected PC. Other reports indicate that Brown Orifice allows files to be displayed and read. In any event, Netscape should have a patch available to thwart Brown Orifice. In the meantime, disable Java in your browser, and you will be fine until a patch can be applied.

Summary and Recommendations

Although this chapter focused on Microsoft's VBS and JavaScript attacks, other types of script attacks prowl the World Wide Web. Common Gateway Interface (CGI) scripts are also making their mark on the Internet community. Recall that CGI is a common program used for providing interactivity, such as data collection and verification in Web servers. Default CGI programs, or scripts, are used to launch a variety of attacks, such as backdoor incursions, credit card theft, and site vandalizing.

In general, disabling embedded scripting eliminates the security risks associated with the entire class of embedded e-mail scripting attacks. More important, most users do not need to have "automatic" scripting enabled, such that double clicking can execute the script files.

With respect to Microsoft applications in particular, disable Windows Scripting Host (WSH) in the operating system or related application. WSH is installed by default with Windows 98 and Internet Explorer version 4.0 and later. Disabling WSH virtually eliminates the possibility of accidentally launching a malicious VBS file. Also disable embedded e-mail scripting capabilities in your e-mail programs and scripting support in browsers. Disabling browser support for scripting virtually eliminates the possibility of getting exploited by a rogue Web site.

For users with legitimate scripting needs, scripts can be executed by using utility applications, such as Wscript.exe program. As a final precautionary measure, remove the .VBS extension from the Registered File Types (Registry) altogether. With this action, you have virtually instituted the most practical security available against malicious code without adding an e-mail or content filter.

One final caveat. In certain applications, disabling embedded scripting support does not eliminate the applications' vulnerability to script attacks. For example, Internet Explorer (IE) versions 4.0 service pack 2 (SP2) or later enables Microsoft's Access files (.mdb files) to be accessed for execution of malicious code. Fortunately, Microsoft provides a patch that fixes this vulnerability. It's imperative that you make it a regular practice to visit the Web site and information repositories that disseminate information on the software programs that comprise your network's applications. Staying in touch on a regular basis will ensure that you obtain and administer the appropriate patches to your network's applications in a timely fashion.

The recommendations here provide the key safeguards to thwarting script attacks. However, for the most effective level of protection, an e-mail filtering or content management system should also be added to your security arsenal. In between applying updates, specific fixes or patches, and general policing to check the status of the network's active scripting support, a content management/e-mail filtering system quarantines or prevents script code from passing through to the enterprise's network and generally provides another layer of protection. In other words, it can actively enforce the corporate security policy against embedded scripting activity by providing an automated level of protection that complements the practical measures that are manually administered in an ongoing basis.

CHAPTER EIGHT

Countermeasures and Attack Prevention

Nearly all enterprises have disaster-recovery plans for their mission-critical systems. Ironically, though, many companies do not have an attack-survival or prevention plan for those very same systems. This is indeed unwise, especially in light of the mad dash toward e-business. Going online with the enterprise network without an attack-survival and prevention plan is like an airport operating without plans for coping with hijacking.

This chapter focuses on how to survive and thwart network attacks. Attacks occur even with security measures, such as firewalls and strong authentication systems, in place. If, for example, the firewall is improperly configured or has ports that are not protected or if passwords in user authentication systems are stolen, guessed, or cracked, you are vulnerable to attacks. This chapter helps you prepare for an attack, covering how to assemble an incident response team, form an alliance with your ISP, report the attack to the appropriate authorities, and collect forensics for legal prosecution.

This chapter also explores another critical area: recognizing a DDoS attack in progress and implementing countermeasures to mitigate its impact. Several proven techniques can be implemented by your security team as practical security measures for the enterprise network. Such measures may be adequate and provide a cost-effective alternative to deploying more expensive third-party or COTS (commercial off-the-shelf) security measures.

Surviving an Attack

External attacks from hackers are like guerilla warfare because the conflict takes place behind your network's perimeter. When the attack originates from disgruntled employees turned internal saboteurs the theater of conflict is again within your network. Attacks from external sources include backdoor incursions, Trojan horses arriving as e-mail, denial-of-service, and related exploits. Through backdoor programs, intruders will engage your resolve for root control of your PC, steal passwords and other sensitive information, destroy and modify data, hijack communication sessions, and plant malicious code. (See the section Backdoor Programs and Trojan Horses in Chapter 6.) DoS attacks succeed in denying service to legitimate users. Exploits through e-mail include slipping backdoor programs into your systems and launching script attacks. On the other hand, internal saboteurs can be employees or employees of business partners. Much of what they do depends on their motivation. Internal attacks include destruction and modification of data, theft and espionage for personal gain and/or profit, and other computer crimes.

Whether the attack originates from outside or within your network, the level of costs you incur when you engage the attack depends, by and large, on the effectiveness of your attack and survival plan. Generally, the more comprehensive the plan, the less costs you incur and the greater your ability to control the effects of the attack. If your attack results in a root compromise, what steps do you take to recover? Or what do you do when fighting a denial-of-service attack? What *tactical* strategies should be deployed when a DoS attack is launched against you? How do you handle internal saboteurs? Before answering these questions, let's begin with some general planning.

Although it is easier said than done, the most important reaction is to stay calm when you are attacked. No question, this will be a stressful time for your company, especially if you deal with a DoS attack. Every moment that service is denied could mean tens of thousands of dollars in lost revenue or productivity. To avert those feeling of helplessness and violation, you must keep your wits about you. The best way to ensure this is by implementing an approved survival plan that is ready to go in the event of an attack.

Formulate an Emergency Response Plan and an Incident Response Team

The emergency response plan is geared toward helping organizations build and improve their attack preparedness. There are no hard-and-fast rules to follow for constructing an effective action plan. However, nothing is more important than assessing the enterprise's readiness or its ability to appropriately respond to a given attack. During this critical *preattack* activity, you will discover the status of business systems, existing security measures, support infrastructure and services—ISPs—and internal resources' predisposition to attack.

Your emergency or incident response plan should cover the following key areas:

- The incident response team (IRT) and its members' roles and responsibilities
- When to review your preattack posture with security checklists
- Where to obtain outside assistance, if necessary
- What law enforcement authorities to call to report the incident
- How to identify and isolate the host(s) under attack
- How to monitor important systems during the attack, using appropriate security tools, such as firewalls, intrusion detection systems, e-mail filters, and/or content management systems or third-party security services
- What tactical steps to take to mitigate the impact of the attack while it is under way

The most important aspect of your plan is the team of individuals selected for responding to an attack and related emergencies. Perhaps individuals also support disaster recovery. Individuals who are assigned to disaster recovery are generally astute in logistics and systems support and function well under pressure. Similarly, to cope with a computer attack, you should assign individuals who also function well under pressure and respond quickly and logically to adverse situations. Additionally, these people must be experienced with network operations. Therefore, your team should be composed of senior technical staff who are capable of formulating and executing a plan of action. Senior management should both sanction and give the necessary authority to the team and the subsequent plan.

Specifically, the team should consist of several key individuals. The most important member is the one who has line responsibility for network security. Ideally, this person should be proficient in the various attack classes and capable of discerning the kind of attack the enterprise sustains. Being able to identify the type of attack will help determine its source, duration, and, perhaps, the number of occurrences. Next, the individual who manages or is proficient in managing computer/security logs is a good candidate. The lead engineer or systems person who implements the network's security measures is also a good candidate for the team. The communication specialist or individual who interfaces with the enterprise's ISP is a sure bet, too. Finally, a team leader should be designated; it doesn't necessarily have to be the security officer but must be someone who can manage the emergency task force effectively.

Granted, some potential candidates may possess two or more of the skill sets that the team requires. If so, the role and the responsibility of a given team member should match the skills he or she is capable of performing best during an attack. In other words, make sure that enough individuals are available to deal with the attack swiftly and effectively. This will prevent any one individual from getting bogged down with too much to do. Figure 8–1 summarizes the baseline organization for an incident response team.

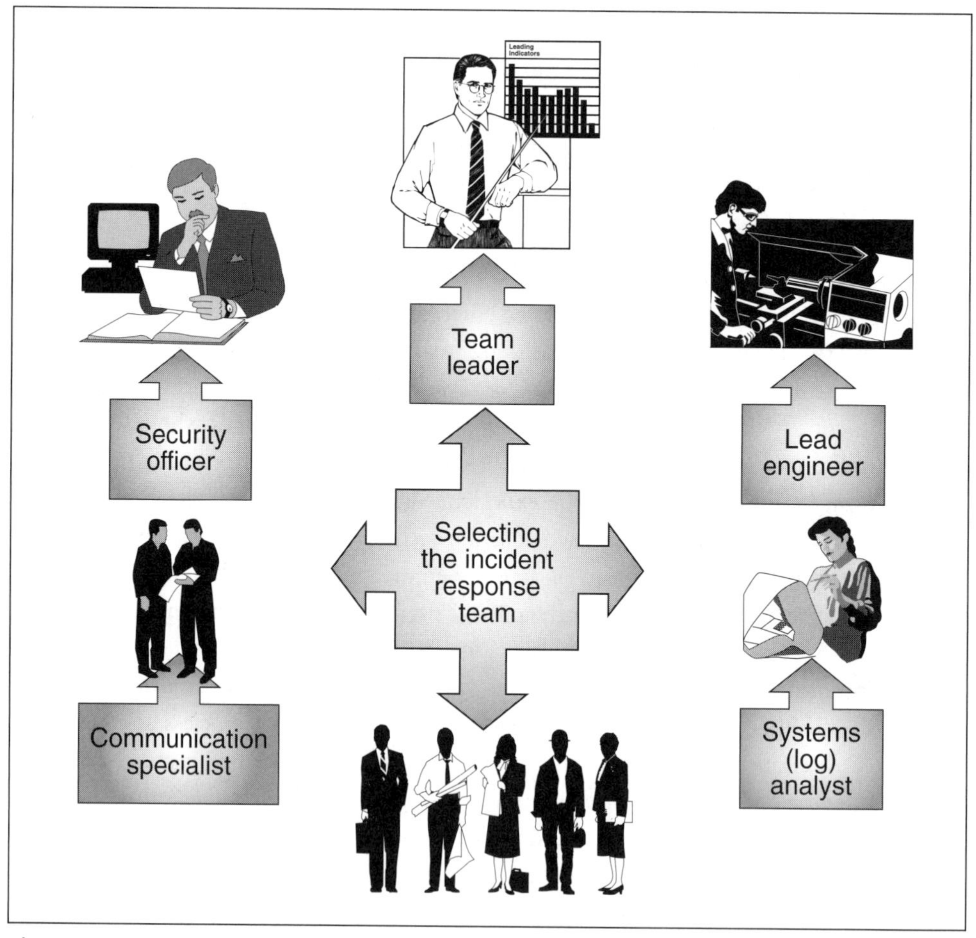

Figure 8–1 Baseline organization for incident response team

If the skills you need are not available in-house, consider outsourcing these requirements to a service provider that offers emergency response services. Companies that provide such services typically offer the means to provide them directly or train staff as needed. If the budget permits, you may also hire the individual(s) you require. However, outsourcing for training or service delivery may be the most feasible choice.

Obtain Outside Assistance

When under attack, get help. Start by informing your Internet service provider that an attack is in process. The ISP can take action to pinpoint where the attack is originating, to block subsequent incursions from reaching the affected hosts in your enterprise. You can also report attacks on your network to the CERT Coordination Center. This is strongly recommended. The CERT/CC is part of the Software Engineering Institute (SEI), a federally funded research and development center at Carnegie-Mellon University. CERT/CC was commissioned to handle and to coordinate Internet emergencies among authorities after the infamous Internet worm incident of 1988, which was responsible for bringing 10 percent of all Internet systems to a halt. Since its inception, CERT has received more than 288,600 e-mail messages and 18,300 hot-line calls reporting computer security incidents or requesting information. From these filings, CERT has handled more than 34,800 computer security incidents, and the CERT/CC's incident-handling practices have been adopted by more than 85 response teams worldwide.

During an incident, the CERT/CC can help the enterprise's IRT identify and correct the vulnerabilities that allowed the incident to occur. CERT has received more than 1,900 vulnerability reports since going into operation. CERT will also coordinate the response with other sites that are affected by the same incursion or incident and interface with law enforcement individuals on behalf of an attacked site. CERT officials work regularly with enterprises to help form IRTs and to provide guidance to newly operating units.

When obtaining outside help, have the following important information handy:

- Company or organization name, telephone number, e-mail address, and time zone
- The host name, domain name, and IP address of the system under attack
- The apparent source of the attack: host name or IP address

- A description of the attack method: back door, DoS; attack tool/type; and related file names, executables, and so on
- Duration or discovery of the attack

The initial information you give CERT is invaluable in gathering forensics for legal proceedings. To use CERT's official form for incident reporting, go to http://www.cert.org/reporting/incident_form.txt. (A sample of this form is given in Appendix B.) At a minimum, you or someone on the IRT should have working knowledge of the various classes of attacks and the ability to diagnose them in the event of such occurrences.

Contact Law Enforcement Authorities

When confronting an attack, make certain that you document and establish an audit trail of your response activity. This will bolster your efforts in accumulating forensics and mitigating the effects of current and, perhaps, subsequent attacks. The importance of documenting every action you take in recovering from an attack is crucial. Recovering from an attack is as time consuming as it is stressful. It is not unusual that very hasty decisions are made under these circumstances. Documenting all the steps you make in recovery will help prevent hasty decisions and will provide a critical record of the steps the enterprise took to recover.

So whom do you call? U.S. sites interested in an investigation should contact the local Federal Bureau of Investigation field office. On May 22, 1998, President Bill Clinton signed *Presidential Directive 63,* which targeted 2003 as the year by which a "reliable, interconnected, and secure information system infrastructure" would be created. Although the FBI was involved with more than 200 cases before the directive, the FBI's caseload quadrupled in the two years after the directive took effect. Those 800 cases ranged from vandalism of Web sites to potential theft of military secrets.

It is imperative that you contact law enforcement officials before attempting to set a trap or to trace an intruder. Any attempts to trap or to trace an intruder may prove fruitless unless you involve a law enforcement agency, which will guide you in the appropriate procedures.

To pursue an investigation, call the local FBI field office. The FBI has 56 field offices with *full-time* computer squads assigned. Be advised, however, that some field offices may not have full-time or any computer incident/crime staff assigned whatsoever. In some cases, the nearest field office with computer crime agents on staff may be a state or more away. For information, consult the local telephone directory or visit the FBI's field offices Web page: http://www.fbi.gov/contact/fo/fo.htm. As an alternative reference source, you can also visit the Web page of the FBI's Washington Field Office Infrastructure Protection and Computer Intrusion Squad (WFO IPCIS): http://www.fbi.gov/programs/pcis/pcis.htm.

The U.S. Secret Service may be the best alternative for certain other incidents:

- Theft or abuse of credit card information: credit card fraud, the illegal confiscation and exchange of credit cards, and blackmail
- E-mail threats on the President of the United States
- Impersonation of the President of the United States through forged e-mail

The U.S. Secret Service can be contacted on its main telephone number: (202) 435-7700. The Secret Service's Financial Crimes Division-Electronic Crimes Section telephone number is (202) 435-5850. In the international arena, contact the local law enforcement agency of the country in question for instructions on how to pursue an investigation.

The federal government has also set up the Federal Computer Incident Response Capability (FedCIRC), an incident response organization for federal civilian agencies. Although the day-to-day operations of FedCIRC are handled by CERT, the General Services Administration (GSA) manages the organization. The following sites and telephone numbers have been established for incident reporting.

- For more information on FedCIRC, go to http://www.fedcirc.gov/.
- Send e-mail to fedcirc-info@fedcirc.gov; call the FedCIRC Management Center at (202) 708-5060.
- To report an incident, civilian agencies should send e-mail to fedcirc@fedcirc.gov or call the FedCIRC hotline at (888) 282-0870.

Use Intrusion Detection System Software

When your enterprise is attacked, isolate and monitor affected systems, using intrusion detection system (IDS) software or related services. Such software simplifies this crucial step, especially if you have both host and network intrusion detection software. Host-based IDS is advantageous in isolating internal saboteurs by tracking and responding to violations of security policy or business rules by internal users. When a host IDS encounters an internal breach or an incursion, network security staff are instantly alerted by e-mail, paging, or related notification. For large networks with many mission-critical servers, host-based intrusion detection can be very expensive. If the cost is prohibitive, find the tradeoff that will enable you to protect mission-critical or strategic hosts in the enterprise network. If host-based IDS is simply out of the question, you may be able to justify a network-based IDS deployment to monitor attack signatures as they are encountered on the network. Then appropriate action can be taken, including instructing the firewall not to accept any more IP packets with attack signatures from the source of the attack.

Furthermore, an IDS can lead you to taking important countersteps during the attack and in postattack mode, such as installing security patches in network devices, increasing bandwidth, and taking related actions in coping with a DoS attack. An IDS can also help you determine whether a frontal incursion is really a diversion that is masking a more serious attack, wherein the *actual* objective is the complete takeover of your systems.

In place of an IDS or other security measure, you have no choice but to engage in hand-to-hand combat or use manual methods. This assumes that, at a minimum, a rudimentary knowledge of the command structure or signatures of the various classes of attacks can be determined by reading logs, inspecting root directories, or discovering unusual ports logging activity at odd times.

Inspection of directories for tell-tale signs of attack could identify known files or executables that produce backdoor or DoS incursions. Inspecting user and system logs, although potentially time consuming, could show unexpected activities, such as unauthorized access to financial records or confidential enterprise information. Or, you may discover suspicious UDP or TCP session activity over five-digit or uncommon, port numbers (see Table 6–2).

Unless your network has fewer than, say, 50 workstations, reviewing logs, perusing directories, and dealing with an attack in progress could prove to be a time-consuming, tedious process. The sheer amount of information that you must sift through, let alone the knowledge you or a team individual(s) must possess and apply expediently, could prove overwhelming, especially if the team is operating without security countermeasures, such as an IDS. In the final analysis, if you have a network that is larger than perhaps 25–50 users, you should strongly consider deploying an IDS. An IDS security measure will complement, if not enhance, the knowledge of your team, respond to an incursion expediently, and enable you to sift through a lot of information at the packet level and thereby decrease your exposure to security risks.

For an excellent perspective on IDSs, refer to the review "Intrusion Detection Systems (IDSs): Perspective" by Gartner. The review can be obtained at http://www.gartner.com/Display?TechOverview?id=320015. Another excellent review is "NIST Special Publication on Intrusion Detection Systems," available from the NIST Web site: www.nist.gov. Finally, for a survey of commercially available IDSs, go to http://lib-www.lanl.gov/la-pubs/00416750.pdf" and http://www.securityfocus.com. SecurityFocus.com is an IDS-focused site that features news, information, discussions, and tools.

Countering an Attack

Firewalls are instrumental in thwarting attacks that involve IP spoofing and certain DoS exploits, such as a SYN-ACK attack. Virus software blocks viruses of all types. Intrusion detection systems are designed to foil virtually every other class of attacks: backdoor programs, distributed DoS attacks and related tools, IP fragmentation exploits, and, especially, internal sabotage. At a minimum, enterprises that operate in the Internet economy should have either a stateful inspection or proxy firewall or, preferably a combination of the two and virus software. Furthermore, all organizations that operate with their networks *online* should also have some form of IDS, at least in theory. Believe it or not, however, many organizations don't even have a firewall, let alone an IDS. That's unthinkable, given the potentially horrifying security problems of the Internet.

Although the outlay for fully deploying an IDS may be beyond the financial means of many enterprises, it should be a key countermeasure in your security arsenal, especially if open access to information assets is involved. The only other way to ensure security is by disconnecting your network from the Internet and relegating it to your largest storage closet!

If you determine—preferably after a risk assessment—that weathering a security breach is more cost-effective than deploying an IDS, the information presented in this section will be helpful. The scenarios in which you could possibly forgo deploying an IDS are very few, such as having a non-mission-critical FTP server or a Web site operating in the DMZ, networks that support nonessential business functions with e-mail messaging, bulletin boards, or corporate chat rooms. Consequently, if you have only virus software and own or are planning to put in a firewall, the information presented in this section is critical to your ability to protect your network. If, on the other hand, you are planning to install an IDS, this section will greatly enhance your working knowledge and improve your ability to use such a security tool in protecting your network's infrastructure.

With an IDS and a firewall deployed, you are automatically notified of attack signatures prowling your network and violations of business rules. With a host IDS in place, you are also automatically alerted when *internal* users perform any unauthorized action, such as data modification, removal, or deletion. In effect, an IDS levels the playing field. Without an IDS, you are, unfortunately, left with plain old due diligence and a lot of intestinal fortitude. If you are operating without an IDS (actually, hardening your infrastructure is a best practice whether you have an IDS or not), harden operating systems and eliminate extraneous services and default user accounts and passwords throughout the network. (Refer to the discussion earlier in this chapter on what to do before an attack and the section Deploying and Maintaining Complex Layered Client/Server Software in Chapter 5 for other recommended precautions.) Thus, when you determine that systems are compromised, you should first disconnect the compromised system(s) and make a copy of the image of the affected system(s).

Disconnect Compromised Host/System from Your Network

When a network node or host has been compromised, take it off the enterprise network. In most cases, the best way to recover from attacks is by disconnecting the compromised host to remove or to clean affected files, delete rogue programs, and rebuild the system, including restoring hard drives. In a DDoS attack, you can take countermeasures during the attack to lessen its severity and to allow you to continue providing services to legitimate users. When the attack has run its course, you may need to remove the affected host to determine whether any tell-tale forensics captured by the affected host's system logs can be used in any legal proceedings. Tactics you can use to mitigate the impact of DDoS attacks *in progress* will be covered in detail in Chapter 9.

After disconnecting a compromised UNIX system, you may wish to operate in single-user mode to regain control. For NT or Windows 2000 systems, switching to local administrator mode is recommended. Switching to these modes ensures that you have complete control over these machines, because the link to any potential intruder is severed. One word of caution, however: In switching to single or local administrator mode, you run the risk of losing potentially useful information because any processes running at the time of discovery will be lost. Processes that are killed may eliminate the ability to mark an intruder's pathway to, for example, a network sniffer.

By maximizing your efforts to obtain important forensics from the processes being run at the time of the attack, you risk further breaches or deeper penetration before the compromised host can be removed. When you are convinced that you have captured vital information from active processes, removing the machine and switching modes would be more feasible.

Working in single-user mode or local administrator mode for UNIX or NT/Windows 2000, respectively, poses some additional advantages. Doing so prevents intruders and intruder processes from accessing or changing state on the affected host as a recovery process is attempted. Recovering in these modes also eliminates the intruder's ability to undo your steps as you try to recover the compromised machine. Therefore, when you are certain that you have obtained adequate forensics from the processes that were running when the attack(s) occurred, disconnect the compromised machine in favor of a single-user operating mode in recovery.

Copy an Image of the Compromised System(s)

Before starting an analysis of the intrusion, most security experts recommend that you make a copy of your system's mass storage device or hard disk. The purpose of the backup is to provide a snapshot of the file directory of the host(s) in question at the time of the attack. Chances are, you may need to restore the backup to its original "attack" state for reference as the recovery and subsequent investigation proceeds.

Making a backup of your hard drive in UNIX systems is straightforward. With a disk drive of the same capacity as the compromised drive, you can use the dd command to make an exact duplicate. A derivation of dd can also be used if, for example, a Linux system with a SCSI (small computer systems interface) disk is the host in question. With a second SCSI disk of the same make and model, an exact replica of the compromised disk can be achieved. (For more information, refer to your system documentation on making a backup of mass storage media.) NT systems have no inherent equivalent command. However, plenty of third-party system utilities are available. Label, sign, and date the backup, and store it in a safe place to ensure the integrity of the unit.

Analyze the Intrusion

With the compromised system(s) disconnected and the recommended backup(s) completed, you are ready to analyze the intrusion. What is the extent of the breach? Were sensitive information and intellectual secrets confiscated? Were any user passwords captured during the raid? Did the hackers get away with confidential customer information, including credit cards?

Analyzing the intrusion or to answer these questions begins with a thorough review of the security person's equivalence of surveillance tapes: log files and configuration files. Through analysis of these files, your security team will uncover signs and confirmation of intrusions, unauthorized modifications, and maybe obscure configuration weaknesses.

Verify all system binaries and configuration files. The best way to accomplish this is by checking them thoroughly against the manufacturer's distribution media that shipped with the system. Keep in mind that because the operating system itself could also be modified, you should boot up from a trusted system and create a "clean" boot disk that you could use on the compromised system. After the boot disk is created, make certain that it is write-protected to eliminate any potential of its being modified during your analysis.

CERT/CC, which has had considerable experience with such matters, has concluded that system binaries of UNIX and NT systems are a favorite target of intruders for perpetrating their incursions. Trojan horses commonly replace binaries on UNIX systems. The binaries that are typically replaced include

- TELNET, in.telnetd, login, su, ftp, ls, ps, netstat, ifconfig, find, du, df, libc, sync, inetd, and syslogd
- Binaries in the directory /etc/inetd.conf, in critical network and system programs, and in shared object libraries

Trojan horses, used on NT or Windows 2000 systems as well, usually introduce computer viruses or remote administration programs, such as Back Orifice and Netbus. (See Chapter 6.) In cases on NT systems, a Trojan horse replaced the system file that controls Internet links.

One final caveat about evaluating binaries. Some Trojan horse programs could have the same timestamps as the original binaries and provide the correct sum values. Therefore, in UNIX systems, use the cmp command to make a direct comparison of the suspect binaries and the original distribution media. Alternatively, many vendors supply for their distribution binaries MD5 checksums, which can be obtained for UNIX, NT and Windows 2000 systems. Taking the checksums of the suspect binaries, compare them against a list of the checksums made from the good binaries supplied by the vendors. If the checksums don't match, you know that the binaries have been modified or replaced.

After evaluating system binaries, check system configuration files. For UNIX systems, make sure to do the following.

- Check your /etc/passwd file for entries that do not belong.
- Check to ascertain that /etc/inetd.conf has been changed.
- If r-commands, such as rlogin, rsh, and rexec, are allowed, determine whether anything doesn't belong in /etc/hosts.equiv or in any .rhosts files.
- Check for new SUID and SGID files. Use the appropriate command to print out SUID and SGID files within the compromised file system.

For NT systems, be certain to check the following.

- Look for odd users or group memberships.
- Check for any changes to registry entries that start programs at log-on or at the start of services.
- Check for unauthorized hidden sharing facilitated by the net share command or Server Manager utility.
- Check for unidentifiable processes, using the pulist.exe tool from either NT's resource kit or Task Manager.

Recognizing What the Intruder Leaves Behind

After ascertaining what intruders have done at the root and system levels, figure out what attack tools and data the hackers left behind. The class of attack that you find will provide insight into the type of breach that is ultimately sustained by the enterprise. Following are the common classes of attack tools, related data, and the enterprise information assets, capital, and other things the intruders were most likely after.

Network sniffers are software utilities that monitor and log network activity to a file. Intruders use network sniffers to capture user names and password data transmitted in clear text over a network. Sniffers are more common on UNIX systems. The equivalent on NT systems are keystroke-logging logs.

Trojan horse programs appear to perform an acceptable function but in reality perform a clandestine exploit. (See the section Backdoor Programs and Trojan Horses in Chapter 6.) Intruders use Trojan horses to hide surreptitious activities and to create back doors to enable future access. Like sniffers, Trojan horses are also used to capture user name and password data. In UNIX system exploits for example, hackers replace common system binaries with Trojan horses or their altered versions of these programs. (See the section Analyze the Intrusion in this chapter.)

Backdoor programs are designed to hide themselves inside a compromised host. (Refer to the section Backdoor Program Functions in Chapter 6.) Typically, the backdoor program enables the hacker to access the targeted system by circumventing normal authorization procedures. The real strength of back doors is that once they are in place, the intruder can access your system without having to deal with the challenges of vulnerability exploitation for repeat access.

Vulnerability exploits are used to gain unauthorized access in order to plant intruder tools in the compromised host. These tools are often left behind in masked, obscure, or hidden directories. For example, intruders exploit vulnerabilities in RPC services to install replacement programs, or Trojan horses. Replacement programs hide intruder processes from system administrators and users. (See the sections Deploying and Maintaining Complex Layered Client/Server Software in Chapter 5 and Attacks through Remote Procedure Call Services in Chapter 7.)

Other intruder tools are the same as those security managers and system administrators use for legitimate operations. Intruders use these tools to probe or scan your network, cover their tracks, replace legitimate system binaries, and exploit other systems while operating from a compromised one.

Basic scanning/probing tools become objects of suspicion when they appear in your network in unauthorized locations, such as directories. Chances are that intruders are using the host in question to scan other networks, and your network has become a base camp for illicit intruder operations. Intruders typically use scanning tools to determine where they will set up shop. Some of the tools scan for a specific exploit, such as ftp-scan.c. If a single port or vulnerability is being probed on your network, the intruder is perhaps using one of these single-purpose tools. On the other hand, such tools as sscan and nmap probe for a variety of weaknesses or vulnerabilities. Of the two, sscan is the most sophisticated multipurpose scanner in the wild. To launch sscan, the hacker simply supplies the program with the network IP address and network mask, and the program does the rest. The hacker must be at the root level to use sscan.

The port scanner of choice since its introduction, nmap, short for network mapper, is a utility for port scanning large networks, although a single host can be scanned. This utility provides operating system detection, scans both UDP and TCP ports, assembles packets in a variety of ways, randomizes port scanning, bypasses firewalls, and undertakes stealth scanning, which does not show up in your logs.

Replacement utilities use utility programs called rootkits to conceal their presence in compromised systems. One of the most common rootkits is Irk4, an automated script that when executed, replaces a variety of critical intruder files, thus hiding the intruder's actions in seconds.

Password-cracking tools, such as Crack and John the Ripper (JtR), are password-cracking tools in widespread use in the wild. Unless you are using these tools for your own benefit, you won't necessarily see them on your network, unless your network is used as a base camp. Crack is freeware and is designed to identify UNIX DES-encrypted passwords through ordinary guessing techniques supplied by an online dictionary. JtR, faster and more effective than Crack, runs on multiple platforms, including Windows operating systems and most flavors of UNIX. *Hacker accounts* on UNIX systems may be called moof, rewt, crak0, and w0rm.

DoS attack tools are likely to be found in files named master.c and/or ns.c, indicating that your network contains either a master or a zombie, respectively, component of a Trin00 DDoS attack system.

Intruder tool output to system logs is cleaned to cover the hackers' tracks. Cloak, zap2, and clean are log-cleaning tools that intruders may use for this purpose. Nevertheless, it is still a good idea to check the logs for intruder activity. For example, syslogd in some versions of UNIX, such as Red Hat Linux, logs sscan probes against your network in the secure log file. Secure log appends an entry whenever an outside host connects with a service running through inetd. However, sscan always interrupts the connection with a reset before the source of the probe can be logged, thus keeping the source of the probe unknown. Logs that contain no information but should are a sign that they have been cleaned, which in turn suggests the presence of a compromised network host.

Unfortunately, the tools described here are only some of them. You need to continue your research into the types and functions of the tools that are used against you, so that the appropriate countermeasures can be deployed to safeguard your network. At a minimum, you should download a copy of nmap and familiarize yourself with this dynamic probing tool. Periodically run scans against your network to see what the hacker sees and what could potentially be exploited. Adopting this practice will help you stay focused on eliminating the vulnerabilities that expose network services to hacker incursions. To download a copy of nmap, go to http://www.insecure.org/nmap/index.html.

In general, look for ASCII files in the /dev directory on UNIX systems. When hackers replace some system binaries with Trojan horses, the resulting *Trojan binaries* rely on configuration files found in /dev. Search carefully for hidden files or directories, which usually accompany creation of new accounts or home directories. Hidden directories tend to have strange file names, such as "." (one dot) or ". " (one dot and a *space*). Once a hidden file is isolated, list the files in that directory to determine the extent of the incursion in terms of hacker tools and files that are on the compromised machine. On Windows systems, look for files and directories that are named to directly or closely match a Trojan system file, such as EXPLORE.EXE or UMGR32.EXE.

CHAPTER NINE

Denial-of-Service Attacks

Denying service to legitimate network users is a favorite exploit of hackers. Hackers may deny access to an individual network service, such as FTP or HTTP, or to an entire network. DoS, especially DDoS, attacks can potentially be the most costly of attacks for enterprises to handle. In large B2B networks, for example, a DDoS attack could cause losses in literally millions of dollars. In addition to incurring internal costs for fighting a DDoS attack, thousands—perhaps even millions—of dollars more could be realized in lost revenue. Most disturbing, once a DDoS attack is launched, you can't stop the flood of packets, mainly because the attack source could be thousands of network hosts sending bogus information from anywhere in the wild. Although you may not be able to stop a DDoS, you might be able to stem the tide. This chapter focuses on denial-of-service attacks and on the countermeasures you should deploy to mitigate the potentially devastating effects.

Effects of DoS and DDoS Attacks

Hackers have used *simple* DoS attacks, such as SYN-ACK and land incursions, in the wild for some time. "Simple" attacks originate from a single or limited source and are launched against a single target at a time. In contrast, a *distributed* DoS is launched from perhaps thousands of locations against a *single* or limited number of targets.

Firewalls protect against simple DoS attacks by, for instance, not allowing half-open, or SYN-ACK, connections to remain on the server in question such that a backlog of SYN-ACK messages occurs. A half-open session exists when the SYN-ACK, or acknowledgment, of the server system has not received the *final* ACK from the initiating client. In this situation, firewalls will send the final ACK to complete the half-open transmission. The server system will reset for subsequent session establishment. Numerous SYN-RECEIVED entries in the firewall logs could suggest a SYN-ACK attack. (See the section SYN-Flood and Land Attacks in Chapter 6.) SYN-ACK and land attacks are also thwarted by the anti–IP spoofing capability of firewalls.

Ping of death and other variants, such as Teardrop (see Chapter 6) are used to send malformed and/or oversized packets to systems. When they receive such packets, hosts operate erratically and possibly crash, denying service to legitimate users. Correctly configured firewalls can recognize these attacks and block them from entering the network because they are originating from a single or a limited source. (Note that in this scenario, the firewall is only repelling the attack, not stopping it from occurring.) Furthermore, if the targeted site has high-speed/bandwidth Internet access, such as a digital subscriber line (DSL), fractional T1 or T1, a simple, or regular, DoS attack will most likely not cause a system crash.

Distributed DoS (DDoS) attacks, however, are another story. These attacks cannot be so easily repelled, even with a firewall. However, with a *fortified* firewall—one equipped with performance boosters—an IDS, and maximum communication bandwidth(s), you may be able to prevent the DDoS from dramatically impacting network performance or crashing the network.

A DDoS attack directs traffic from hundreds to thousands of systems against one or a limited number of targets all at once, creating an enormous flood of traffic at the victim's network/host(s). Like simple DoS attacks, DDoS attacks arrive as a SYN-ACK attack, Ping of death, UDP bomb (ICMP Port Unreachable attack), or a Smurf (ICMP bandwidth attack). (See Chapter 6 and Appendix D.)

A DDoS attack is a war of bandwidth, with the attacker most likely to win because of the potential for incredible numbers constituting the attack source. Your challenge involves enduring an attack by minimizing the strain on computing resources, thereby preventing the complete consumption of connectivity bandwidth and recovery.

During a DDoS attack, a firewall bears the brunt of the attack. Any firewall, if configured correctly, could recognize all the aforementioned DDoS attack methods and drop the packets before they penetrate the network. However, if the firewall's rule base is not properly configured or if the server lacks sufficient computing cycles, it will quickly become consumed with processing bogus connection attempts. Even a properly configured firewall would still need time to make decisions before dropping the connection attempt. Processing a flood could strain the firewall's resources, leading to performance degradation and denial to any *legitimate* connection attempts.

To determine whether an attack is indeed a DDoS attack, you should consider the MAC (media access control) addresses being logged by the firewall. Looking for source IP addresses would not necessarily help, because they are likely to be spoofed. Further, DDoS attack tools use *flat-distribution random-number generators* that cause each attack packet to use an address only once. Typically, these addresses are nonroutable IP addresses or ones that don't exist on global routing tables. Moreover, the MAC address is the permanent address assigned to devices that provide the interface, or link, for LAN (local area network) communication and data access. For example, the manufacturers of network interface cards (NICs) usually burn in the MAC address into PROM (programmable read-only memory) or EEPROM (erasable programmable read-only memory) configured into the LAN card. If you see multiple attacks from various MAC addresses, you are probably sustaining a DDoS assault.

General Computing Resources

As a practical measure, make sure that your firewall gateway/server has ample computing resources—RAM (random-access memory), CPU (central processing unit) cycles, and caching—that are over and above the recommended requirements. Expanding RAM could enable the operating system to continue functioning, perhaps even maintain performance, during the attack. The extra capacity may also enable your firewall to address the attack packets while allowing existing traffic to continue with their sessions. (Note that although *existing* traffic at the time of the DDoS attack may not be denied, session requests from *new* traffic might be.)

High-Performance Firewall

Implementing a high performing firewall entails deploying two firewalls with at least two load-balancing devices to distribute traffic equally among the resulting firewall cluster. Load balancing, provided by specialty hardware, is a technique that maximizes resource availability and performance in networks. The relatively inexpensive cost of load-balancing solutions makes them a cost-effective alternative for optimizing security resources.

Load-balanced firewalls increase performance benefits by allowing both firewalls to make security decisions simultaneously. In a proxy-based firewall cluster, for example, each firewall could support up to 850,000 concurrent sessions. Although a clustered, load-balanced firewall solution will not *stop* the DDoS attack, the firewall cluster will lessen the impact of the attack by marshalling the resources of two firewalls simultaneously.

In a load-balanced firewall implementation, all firewalls are active and continually sharing state information to determine one another's ability to control network access. In other words, the firewalls are continually checking one another's ability to perform the job. During a DDoS attack, it is inconceivable that a firewall cluster could be overwhelmed to the point of crashing, especially as each firewall in a load-balance farm could handle up to 850,000 concurrent sessions, depending on the hardware configuration.

However, depending on the magnitude of the attack, even a load-balancing firewall solution could, potentially, struggle to keep performance from degrading or bandwidth from completely exhausting in a DDoS attack that is generating millions of packets attempting to establish bogus TCP or UDP sessions with your network. But your ability to stay online and to keep network resources accessible is increased considerably with a firewall cluster as you work with your response team and ISP to put down the attack at the source.

Network Bandwidth

The impact of a DDoS incursion can be lessened by taking precautions with the bandwidth of critical network implementations. For example, upgrading from a fractional to a full T1 and distributing processing across the network may help achieve a brute-force bandwidth defense. Such defense combines large communication channels with distributed networks to provide the brute strength capable of alleviating potential bandwidth constraints in the event of a DDoS attack.

One way of achieving distributed networks is through implementing a load-balanced distributed network with high availability. Such a network incorporates *server farms,* with redundant applications to optimize performance and network availability. Load-balanced server farms allow user sessions to be balanced among them. If any server goes down, user sessions are automatically balanced among the remaining ones in operation, without any interruption of service. Although it will not impede the DDoS attack, the extra bandwidth may minimize the potential for a bottleneck at the point of entry and exit from the network, as the fortified resources of the distributed networks and large communication pipes handle the bogus traffic.

Handling a SYN Flood DDoS Attack

A SYN packet is the first one sent during a TCP session setup. FTP, TELNET, and HTTP services rely on TCP to establish connections and to relay messages. By sending only a SYN packet and no subsequent packets in response to a SYN-ACK, or acknowledgment, from a server, the TCP session request is left half open: an orphan TCP session.

When a DDoS attack system, such as Stacheldraht or Tribe Flood Network 2000 (TFN2K) dispenses a flood to one or a range of IP addresses, the servers or hosts targeted at those addresses receive numerous SYN packets. These attacks randomly target ports, such as FTP and HTTP, at the target machines. The objective of the SYN flood is to disable the target machine(s) by opening all available connections, thereby perhaps crashing the system or denying legitimate users from accessing the server.

A SYN flood is difficult to detect because each open session looks like a regular user at the Web or FTP server. The extent of the flood damage depends on how the source addresses are spoofed. SYN flood packets can be spoofed with either *unreachable* source IP addresses—addresses that don't appear on global routing tables—or *valid* IP addresses. When hackers launch attacks using IP source addresses created by a random-number generator or an algorithm that allows IP source addresses to be changed automatically, the source address is unreachable. When spoofed source addresses are unreachable, only the target system is affected. The targeted host server repeatedly reserves resources, waiting for responses that never come. This continues until all host resources have been exhausted.

In contrast, when a SYN flood is launched with spoofed IP source addresses that are valid or legitimately reachable, two systems are affected: the target system and the network that is assigned the valid IP addresses. The network that owns the valid IP addresses is a collateral, or unintended, victim when the attacker spoofs the SYN flood packets with that network's IP addresses. Consequently, when SYN packets are spoofed with valid IP addresses, the destination target system forwards SYN-ACK responses to the network that it believes to be the originator of the TCP session requests. The network assigned to the valid IP addresses receives the SYN-ACK responses, although it never initiated the TCP sessions or sent the original SYN packets. When this occurs, the unwitting hosts are forced to use resources to handle the flood of SYN-ACK responses that they did not expect. In this scenario, the hacker succeeds in degrading performance or crashing networks in two separate network domains.

Countermeasures

IP source address spoofing is a fact of life in SYN flood DDoS attacks. The hackers want to cover their tracks and mask or conceal their true whereabouts or base of operations. IP address spoofing is a favorite tactic for masking assaults against unsuspecting targets. When packets with fake source addresses—whether randomly generated or valid for another network—originate from your network, you become an unwitting accomplice to hacker offenses. To prevent packets with spoofed source addresses from leaving your network and to eliminate the chance of becoming an unknowing participant in hacker capers, consider implementing a technique called *egress filtering.*

The opposite of *ingress filtering,* egress filtering says that on *output* from a given network, deny or don't forward data that doesn't match a certain criterion. With egress filtering, network managers can prevent their networks from being the source of spoofed IP packets. This type of filter should be implemented on your firewall and routers by configuring these devices to forward only those packets with IP addresses that have been assigned to your network. Egress filtering should especially be implemented at the external connections to your Internet or upstream provider, typically your ISP. The following statement is an example of a typical egress filter:

```
Permit Your Sites Valid Source Addresses to the Internet
Deny Any Other Source Addresses
```

All enterprises connected to untrustworthy networks, such as the Internet, should ensure that packets are allowed to leave their networks only with valid source IP addresses that belong to their networks. This will virtually eliminate the potential of those networks' being the source of a DDoS attack that incorporates spoofed packets.

Unfortunately, egress filtering will *not* prevent your network from being compromised by intruders or being used in a DDoS attack if your network's valid addresses are used in the assault. Nevertheless, egress filtering is an effective countermeasure and a potentially effective security practice for limiting attacks originating with spoofed packets from inside enterprise networks. In general, it's everyone's responsibility to institute best practices, such as egress filtering. Only with these types of measures will enterprises ultimately achieve a reasonable level of security when operating in the Internet environment.

Precautions

Because a SYN DDoS flood is so difficult to detect, most vendors have fortified their respective operating systems to be resilient in the face of an attack. Operating systems have been modified to sustain attacks at very high connection attempt rates. Find out from your vendors what their threshold is for the maximum rate of connection attempts of each operating system that your enterprise has deployed in e-business systems. To achieve maximum limits may require upgrading software releases.

Load-balancing devices also incorporate mechanisms for eliminating orphan SYN connections. Even if egress filtering is implemented at the router and operating systems are fortified at the servers, load-balancing devices can offer additional security countermeasures and increased bandwidth. Load-balancing devices, such as those provided by Radware, eliminate orphan SYN sessions after 5 seconds if no additional SYN packets are received at the server host and within their own session tables. Combining egress filtering, fortified operating systems, and load balancing creates an effective countermeasure for SYN DDoS flood, especially when the source address is spoofed.

Handling a Bandwidth DDoS Attack

Every DDoS attack is a war of bandwidth. The Smurf attack, however, could be the most devastating because the resources of three networks—those of an accomplice, an intermediary, and a victim—are typically commandeered for the exploit. (For a full description of the Smurf attack, see the section ICMP Directed Broadcast, or Smurf Bandwidth Attack, in Chapter 6.)

A Smurf bandwidth attack is usually leveled against targets from a DDoS attack system, such as Tribe Flood Network (TFN), TFN 2000, or Stacheldraht (see Table 6–1). Attackers orchestrate hundreds or thousands of compromised hosts to simultaneously send ICMP echo request or normal ping packets to the IP broadcast address of a target network. That address is designed to broadcast the packet to every network host that is configured for receiving the broadcast packet. At this point, the network that receives an IP directed broadcast attack of

ICMP echo request packets becomes the *intermediary* in the Smurf attack. On receipt of the packets, all hosts on the network respond by sending ICMP echo reply packets back to the source address contained in the broadcast of echo request packets.

As you might guess, the source address used in the Smurf exploit is spoofed, usually with valid or routable IP addresses of the intended target instead of invalid or unreachable IP addresses. When the intermediary or amplifier network responds with ICMP echo reply packets, the replies bombard the network that owns the spoofed source address. Consequently, this network becomes the ultimate victim of the Smurf attack.

The Smurf attack can be devastating because each ICMP echo request packet that originates from hundreds of accomplice networks compromised by a DDoS system will be responded to in turn with successive waves of ICMP echo reply packets. Each wave is formed by a simultaneous response of echo reply packets from all the hosts on one or more intermediary networks. Thus, if an intermediary network receives hundreds or thousands of ICMP echo request packets, a corresponding number of waves of echo reply packets might respond.

In the Smurf incursion, the intermediary network is also a victim because it is an unwilling participant of the attack, and its resources—bandwidth—will also be consumed, responding with waves of echo reply packets to the victim's network. The bottom line is that both the intermediary and the victim's networks can experience severe network congestion or crashes in a Smurf attack. Likewise, even small to medium-size ISPs providing upstream service can experience performance degradation that will effect all peer users relying on the ISP's network.

Guarding against Being an Accomplice Network

An accomplice network is one that has been breached and enslaved by a zombie, or DDoS daemon, which is controlled by a master server safely tucked away in the wild. A Smurf attack is orchestrated through multiple master servers, which collectively direct hundreds of daemons to send ICMP echo request packets with spoofed source addresses to one or more intermediary networks.

The most effective way to prevent your network from being an unwitting accomplice in the Smurf incursion is by implementing egress filters at the firewall and on routers, especially on external routers to your ISP. With egress filters in place, any packet that contains a source address from a different network will not be allowed to leave your network.

Note that egress filters will not prevent hosts on your network from being compromised and enslaved by zombie DDoS daemons. Most likely, the attacker exploited a network vulnerability to secretly insert the daemons in the first place. (Preventing attackers from exploiting your network to install the various components of a DDoS attack tool is discussed later in this chapter.)

Guarding against Becoming an Intermediary Network

If your network can't be compromised for either an accomplice or an intermediary, the Smurf attack can't happen. That's the good news. The bad news, however, is that too many networks operating in the wild haven't been equipped with the proper precautions. You can institute several steps both externally and internally to prevent your network from being used as an intermediary in a Smurf attack.

For external protection, one solution involves disabling IP directed broadcasts on all applicable network systems, such as routers, workstations, and servers. In almost all cases, this functionality is not needed. By disabling—configuring—your network routers to not receive and forward directed broadcasts from other networks, you prevent your network from being an intermediary or amplification site in a Smurf attack. Detailed information on how to disable IP directed broadcasts at Bay or Cisco routers, for example, are available directly from those manufacturers.

Some operating systems, such as NT—Service Pack 4 and higher—disable the machine from responding to ICMP echo request packets sent to the IP directed broadcast address of your network. This feature is set by default in NT. As a rule, all systems should have directed broadcast functionality disabled by default. Check with your vendor to determine the status of this feature for your operating system and related systems. In fact, make certain that your security policy ensures that this feature is always disabled for systems deployed in your enterprise.

As for internal threats, if a host on your network is compromised and the attacker has the ability to operate from within your network, he or she can also use your network as an intermediary by attacking the IP broadcast address from within the local network. To eliminate the internal threat, make certain that your operating system and related systems are modified to reject packets from the local IP directed broadcast address.

Guarding against Being a Victim

If you are the victim or the recipient of the ICMP echo reply packets from one or more intermediary networks, your network could be placed in a quandary, even if you disable IP directed broadcasts at your router and related systems. The problem is that considerable congestion between external routers and your upstream ISP provider would still exist. Although the broadcast traffic would not enter your network, the channel between your network and the ISP could experience degraded performance and perhaps restrict access to network resources. In this situation, work with your ISP to block this traffic within the source of the ISP's network. (See the section Obtain Outside Assistance in Chapter 8.)

You should also contact the intermediary in the attack once you and your ISP pinpoint the origin of the attack. Make certain that the ISP implements the recommendations provided in the previous section. You might also assist your ISP in identifying the origin of the attack. Several freeware software tools will enable you to accomplish this. The most well known such utility is Whois, a UNIX command that allows victims to obtain contact information on an attacking site. Whois is run from a TELNET window to pinpoint the attacker by running an inquiry on the attacker's IP address against a Whois server. For more information on how to use the Whois command, go to ftp://ftp.cert.org/pub/whois_how_to.

Handling a UDP Flood Bomb

In a UDP flood, the attacking system, controlled by a DDoS attack tool, dispenses a large number of UDP packets to random destination ports on the victim's system. Favorite targets of UDP attacks are the diagnostic ports of the targeted host. A UDP flood, or bomb, causes an explosion of ICMP Port Unreachable messages to be processed at the targeted machine. The affected host gets bogged down because of the enormous number of UDP packets being processed. At this point, little or no network bandwidth remains. Performance degrades considerably, often resulting in a network crash.

One of its more insidious features is that a UDP flood cannot be reliably distinguished from a UDP port scan, because it is not possible to determine whether the resulting ICMP messages are being monitored. Using a scanning tool, such as nmap, an attacker scans a network's host systems to determine which UDP ports are opened. Knowing what ports are open, the attacker knows which ports to attack. At this point, however, it is difficult to determine, on the network side, whether it is a scan or an attack. Nevertheless, if your firewall logs show, for example, approximately ten UDP packets that have the *same* source and destination IP addresses and the same source port but different destination ports, you likely have a UDP scan.

In contrast, a UDP flood can be detected. The first method involves identifying numerous UDP packets with the same source port and different destination ports. Second, as in all distributed DoS attacks, you will see a number of packets with different source IP addresses. If, for any given source address, you find approximately ten UDP packets with the same source port and the same destination address but different destination port numbers, you could have a UDP flood. Another method entails looking for a number of ICMP Port Unreachable messages with multiple varying source addresses but the same destination IP addresses. This method could also signal a UDP bomb from a DDoS incursion.

One method of thwarting a UDP flood is to never allow UDP packets destined for system diagnostic ports to reach host systems from outside their administrative domain. In other words, deny to internal hosts any UDP packets that originate from external connections. One method of denying entry of UDP packets is to disable UDP services at the router. This is usually accomplished with a simple statement, such as "No service UDP (specific to Cisco routers)." Determine how this service can be disabled at your particular routers, and act accordingly.

To determine what ports are open, the attacker relies on port scans and the target host's returning ICMP Port Unreachable error messages. With this information, the attacker knows what ports to attack. To thwart UDP scans altogether or, ultimately, UDP floods, disable the router's ability to return Port Unreachable messages. If necessary, check with your router manufacturer to ascertain the required command(s) to disable this facility at each external interface.

Using an IDS

Fighting DDoS attacks is a community effort. If you are operating in the Internet community, you should take every precaution economically feasible for your enterprise. Therefore, you should consider an intrusion detection system (IDS) to fight DDoS incursions and other security threats.

A DDoS attack system requires several stages to get in place. First, a potential attacker must scan your firewall, assuming that you have one, to determine what ports are open on your network. This is usually accomplished with a scanning tool, such as nmap, SATAN, or Nessus. (Nessus can be obtained from www.nessus.org.) (For information on how to obtain SATAN, see Chapter 11.) Second, an exploit must be run on a vulnerability to gain access into your network. At this point, the intruder would have to establish a connection to an internal host, typically the one that let him or her in, to transfer the attack tool components: master, daemons, and related binaries. Finally, on completing and fully installing the rogue software, the attacker would search for other vulnerable machines in the same network and repeat the process. If a good IDS were in operation on this network, the attacker would never had advanced beyond the initial scan.

Consequently, a firewall operating in tandem with an IDS is the most expedient way for enterprises to detect and to thwart DDoS activity or attacks in *setup mode* or, especially, *in progress*, respectively. An IDS monitors your network like a watchdog. With an IDS on hand, signatures of installed DDoS attack tools or activity based on communication between master and daemon components could be easily detected during setup attempts. Additionally, some IDSs work in conjunction with commercial scanning tools, such as Symantec's Net Recon or Internet Security System's Internet Scanner, each of which could detect installed daemons, agents, or masters before communications transpire. The particular attack tool that you find on your network might be instrumental in helping you determine where corresponding or companion components exist on external networks. Those sites should be contacted accordingly to dismantle the distributed DoS attack system.

When attacks are in progress, an IDS could reconfigure the firewall automatically to shut down the port and the related service that was exploited by the DDoS incursion and to kill the bogus connections in question. On the one hand, the ability to reconfigure the firewall on the fly will help prevent DDoS packets from entering and slowing down the network. In addition, killing the bogus connections in a SYN flood, for example, will free up some valuable resources and reduce the strain on the firewall. On the other hand, to deny access to a particular service, an intruder would have to spoof only packets destined for that service.

To fix to this problem, the IDS can direct that the first packet from each source address be dropped. The current generation of DDoS tools generates packets with random source addresses, whereby each address is used only once. This method works because in TCP-based sessions, TCP sends a second request to the server or Web site, allowing the next and subsequent packets through to the targeted host. Other protocols, such as UDP and ICMP, can be configured to send a retry after the first rejection, to allow normal packets through.

An effective IDS is the electronic sentinel of your network. In operation, its job is to churn through every packet that traverses the firewall, network segment, or host to identify signatures that exploit network vulnerabilities. A particular vulnerability does not have to exist on your network to encounter related scans and attack signatures. And just because your network doesn't have the particular vulnerability does not mean that you shouldn't worry. In reality, an intruder has succeeded in breaking through your defenses. If the right vulnerability was not found on this incursion, perhaps one or more will be found on the next. When an IDS detects DDoS or any other class of attack signatures, the appropriate actions are taken, including alerting the network police and shutting down the port, or window, that let in the rogue packets. The bottom line is that a good IDS is the electronic watchdog, or that critical layer of security, that may mean the difference between becoming an unwitting accomplice in a DDoS attack group or operating safely while minimizing risks to the Internet community.

Recovering from a DDoS Attack

Depending on which side of the DMZ you are on, recovering from a DDoS attack could entail pursuing one or more distinct courses of action. For example, if either the ISP and/or the attacked enterprise has served you notice that your network has been an unwilling accomplice in a DDoS attack system, certain hosts in your network have been exploited, particularly with either master or zombie agents. Provided with the host or host IP addresses, the machines in question should be immediately removed from the network and steps taken to delete the rogue source code, binaries, and related attack signatures/tools that compromised your machines. A good host vulnerability scanner, such as ISS's System Scanner, may need to be used to detect the signatures of the installed attack components. Given the level of an exploit or the depth of penetration, too many system binaries may have been replaced by Trojan horses or hidden directories discovered to trust the host again, even after corrective measures are taken. Under these circumstances, be prepared to reinstall the operating system and applications from scratch to ensure that the host can be trusted when it is back in service.

If you are the target of a DDoS attack, recovery may involve fortifying your network, based on some of the suggestions given in this chapter. A good number of recommendations have been discussed to increase network bandwidth at critical network interfaces, including the firewall. After an analysis of the DDoS intrusion, determine the tradeoff(s) and implement the measure that mitigates the greatest risks to network bandwidth and performance.

Finally, the related logs—system, firewall, and so on—will provide an account of the attack, which your emergency response team can use to work with your ISP to filter out subsequent attack traffic from bombarding your network. This information will also help the FBI track down the zombie machines—domains—that are launching the attacks against you and perhaps the master machines and attacker. In the final analysis, logs provide important evidence that law enforcement needs for successful prosecution. Therefore, make certain that information provided by the firewall logs and the system logs of the host targeted by the attack is collected and protected before it is accidentally or deliberately erased.

CHAPTER TEN

Creating a Functional Model for E-Security

By now, you know how the intruders attack you and what weapons the attackers use. You also know that their tactics, strategies, and methods are to exploit you in clandestine ways, especially when you least expect it. But the scariest facts about intruders are that: they are relentless, patient, and skillfully use a variety of tools to look for obscure windows and back doors into your network. Another sobering factor is that with your information assets online for your business partners, customers, and/or suppliers, potentially thousands of windows and doors created by the network components and resulting applications comprise your computing system. The challenge and lifeblood of e-business security is to replace, eliminate, lock down, and reinforce those windows and doors in your network.

This chapter looks at how to address that challenge, reviewing the steps you should take to mitigate the myriad classes of network vulnerabilities and other critical exposures that could provide unwanted access to your e-business applications. This chapter details a functional model for e-security, incorporating a layered approach to instituting security measures. This functional model is dependent on an understanding of your business objectives and the related security safeguards required as a business enabler.

At the core of this functional model are policy considerations for the enterprise. What policies and/or business practices are necessary to achieve a secure computing environment? As for the network itself, what steps should be taken at the perimeter of your network? Are the operating systems driving the hosts sufficiently ratcheted down? Is the current network architecture optimized for security, or should other steps be taken? What are they? These and other related questions are addressed in full detail in this chapter. When you are done, you will possess the knowledge required to build an effective security architecture that minimizes your vulnerabilities and exposure while achieving the business results that your enterprise desires.

Developing a Blueprint for E-Security

In many respects, developing a blueprint for e-security is a straightforward process that requires more practical experience than innovation, although innovation is a key commodity and one of the critical factors in keeping a little ahead in the security game. Your enterprise's e-security blueprint should translate into a practical functional model, which in turn should lead to a robust architecture that provides the framework for your enterprise's life-cycle security. As with any architecture, that for e-security contains some basic infrastructure components.

The best approach in establishing an e-security functional model, or *blueprint,* for your e-security architecture is to begin with information security best practices and security policy. The security policy should be a living, active document that reflects the business goals and strategies in the current business cycle. (In Chapter 5, see the sections Reengineering the Enterprise Security Policy and Rigidity of Enterprise Security Policy for a definition and in-depth discussion, respectively.) The security policy delineates who can do what in the enterprise network. The security policy also reflects *how*—what standards should be adhered to—when accessing enterprise information assets. For example, DSL or greater connections may be a mandatory policy for accessing certain applications from home offices to ensure acceptable bandwidth and response times.

Tightly connected with the enterprise's security policy are information security best practices. Information security best practices promote the idea that security should be managed to optimize privacy, integrity, and availability, along with the means to assess, or measure, the effectiveness of security elements. Information security best practices also take into account nonrepudiation and the fostering of trust to instill confidence in both internal and external users in the enterprise computing system. (See the section The E-Security Dilemma: Open Access versus Asset Protection in Chapter 2 for a discussion on nonrepudiation and trust.) Thus, at the heart of the e-security functional model is the enterprise

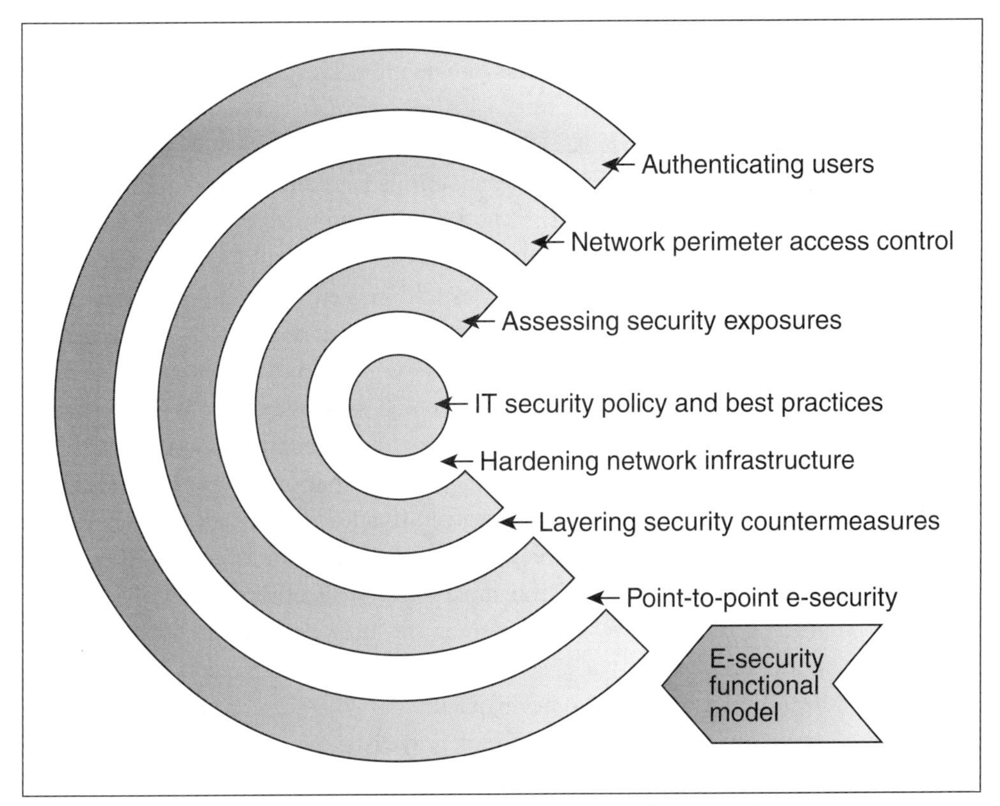

Figure 10–1 Blueprint for success: the e-security functional model

IT security policy, tightly coupled with information security best practices. (For a summary of the recommended e-security blueprint, see Figure 10–1.)

Understanding Business Objectives

Before engaging in a detailed review of each of the components in the e-security blueprint, you need to understand your strategic business objectives. The process typically starts by fully appreciating management's goals for the enterprise. For example, domestic sales for product A must increase by 20 percent. Sales for product B must increase 10 percent per region. On the other hand, support costs must remain constant or, where possible, be reduced. Depending on the organization, accomplishing such goals could be unrealistic. At the least, goals should be challenging; at best, attainable.

Assuming that they are attainable, the next step is putting the information system in place to support the effort. This may require building an extranet to bring the suppliers of products A and B online so that the field can have up-to-the-minute product and order status information. This also may entail setting up a server in a *private* DMZ for internal users—road warriors, mobile users, and remote offices—and in a *public* DMZ for customers and prospects.

After the potential users have been identified and their information platform determined, the focus should shift to the applications. For example, what network services or protocols are required to support the resulting e-business applications? In other words, what services and related ports must be permitted through the firewall to support the strategic applications? HTTP service for the Web application and SMTP for e-mail must be allowed. But what about enabling TELNET? Certain remote offices must TELNET to the server for access.

In such cases, perhaps *ingress filtering* could be used for this group, especially if the IP address is known. Ingress filtering says that on *input*—router or firewall—accept incoming TELNET for a given IP address or IP address range. This rule could be included at the firewall or an access control list (ACL)/filter at the external router. Any TELNET transmissions for all other IP addresses would be summarily dropped by the router or the firewall. TELNET transmissions with spoofed packets or packets with addresses not belonging to the *permitted* IP address or range would also be dropped.

Ingress filters, however, would not prevent spoofed packets that transmit with valid IP addresses. Typically, valid IP addresses are swiped, for example, during a hop through the Internet by an attacker with a sniffer program installed in an unprotected node. Once the valid or routable address of the packets from the permitted TELNET sessions are confiscated, they could be launched from a compromised site and be passed by your router or firewall. Given the potential for this exploit, you must assess the risk to your private DMZ from the remote locations in question. If the risk were still unacceptable, even after installation of an ingress filter, the next step would be to use a VPN (virtual private network). With a VPN equipped with the proper protocol, such as IP Security (IPSec), which is the tunneling protocol standard supported by the Internet Engineering Task Force (IETF), the IP addresses in the header would be encrypted so that the valid IP address could not be deciphered if illegally obtained.

To summarize, gaining a full appreciation of your business objectives and how they translate to IT requirements and, ultimately, e-security safeguards is critically important. Only after you know this critical mapping will you be able to institute the e-security your enterprise requires for fulfilling business objectives. (See the section How E-Security Enables E-Business in Chapter 2.)

Honing in on Your IT Security Policy

One important step should be considered or developed after identifying business objectives but before harnessing computing applications to support them. Specifically, this step entails adhering to the enterprise's *IT* security policy. In general, policy is defined as the rules for obtaining the objectives of an activity, usually a crucial one. The IT security policy, therefore, provides the rules for obtaining the objectives of IT security. And, of course, the objective of IT security is to enable e-business. By definition, the IT security policy must also take into account the business objectives and strategies of the current business cycle. In revisiting our functional e-security model, Figure 10–2 depicts a more applicable portrayal of the e-security functional model and its interdependency with the enterprise's business objectives.

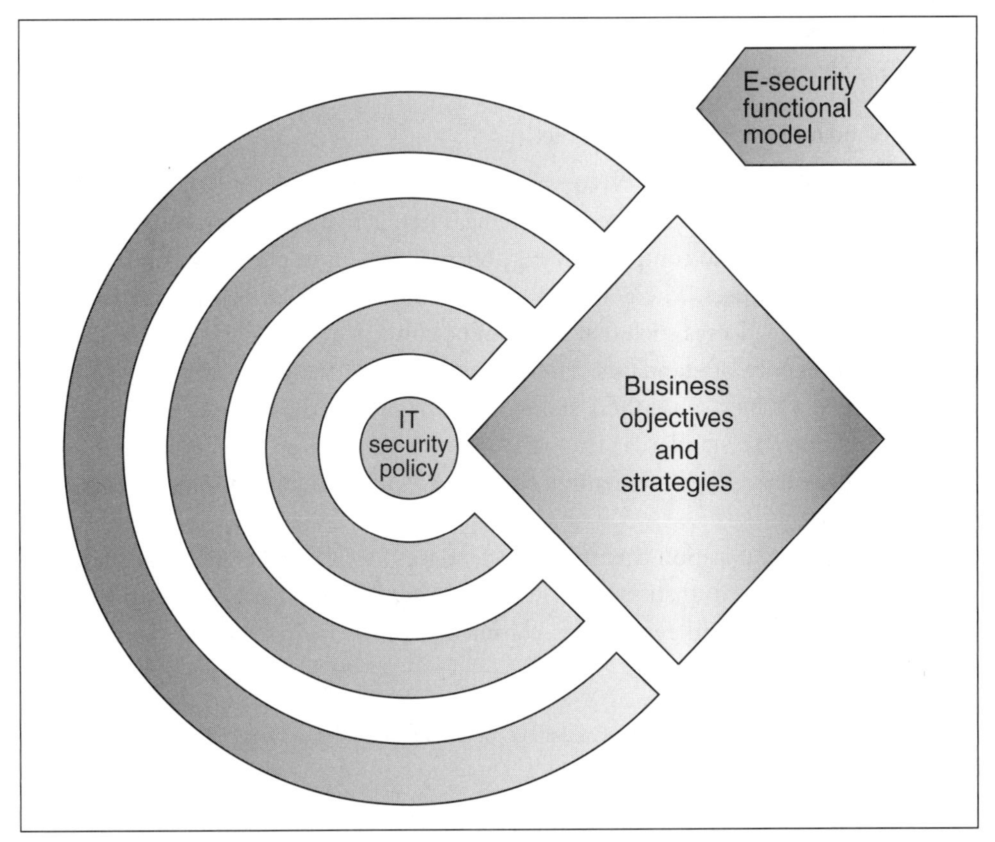

Figure 10–2 Mapping the e-security functional model to business objectives

Note that the IT security policy does not exist under its own auspices and, more important, is not created in a vacuum. On the contrary, the IT security policy should be distilled from the general corporate business policy, corporate security policy, and corporate IT policy. Some enterprises may find it helpful to include the corporate marketing policy as well. Figure 10–3 shows the relationships among these policies.

Interestingly enough, many enterprises do not have a *written* IT security policy. The reason may vary. However, if any of the other policies is not clearly defined or, worse, not written, it's not difficult to infer why many enterprises operate without a written IT security policy. For a comprehensive review of how to build an IT security policy, including recommended elements, refer to an authoritative document produced by the joint collaboration of the International Organization for Standardization (ISO) and the International Electrotechnical Commission (IEC).[1] The bottom line is that the IT security policy and related policies should be written. A written IT security policy is a recommended best practice because in essence, if you can *articulate* it, you can do it.

Making Good on IT Security's Best Practices

If the IT security policy comprises the rules for achieving the objectives of e-security, the *best practices* are the measures, activities, and processes that have been *optimized* to comply with the policy. For the most part, this chapter focuses on best practices. The good news is that the body of knowledge for IT security best practices is such that they can be cultivated and legitimized by the driving influence of standards. The bad news is that the arms race for control of cyberspace may cause these standards to be modified or revised at a faster rate than other IT standards. Thus, the question remains as to whether these standards will ever be sufficiently substantive, applicable, and timely to be widely adopted.

The organization championing standards for IT security best practices is the British Standards Institute (BSI).[2] The BS7799 standard has been fast-tracked for dissemination and request for comments, which invites written commentary,

---

1. "Information Technology—Guidelines for the Management of IT Security Part 2: Managing and Planning IT Security," document ID ISO/IEC TR 13335–2 (E), available from ISO Central Secretariat, Case Postale 56, CH-1211 Geneva 20, Switzerland.

2. The source, or standards document is BS7799–1: 1999, titled "Code of Practice for Information Security Management."

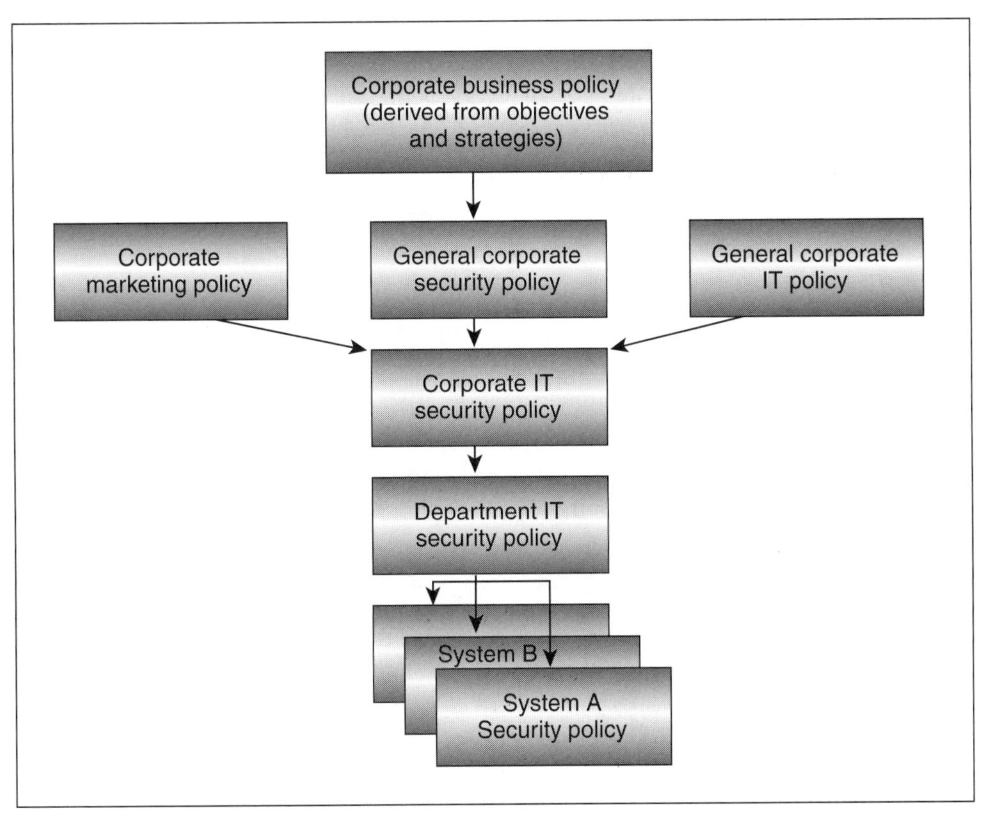

Figure 10–3 The Relationship of IT security policy to general corporate policies (From ISO/IEC TR 13335–2: 1997, Figure 2. Reproduced with permission of the International Organization for Standardization, ISO. This standard can be obtained from the Web site of the ISO secretariat: *www.iso.org*. Copyright remains with ISO.)

suggestions, modifications, endorsements, and, ultimately, adoption from IT professionals worldwide. The ISO/IEC has also endorsed BS7799, accepting it as an international standard in December 2000 as "ISO/IEC 17799:2000 Information Technology—Code of Practice for Information Security Management."

BS7799, a compilation of information security best practices, was developed as a result of industry, government, and commercial demand for a common framework to enable enterprises to develop, implement, and measure effective security management practices and for instilling confidence in e-business commerce. BS7799 is based on the most effective information security practices of leading British and international businesses. The collection of best practices has met with international acclaim for its ability to promote confidentiality, integrity, and availability in e-business networks. The major benefit of BS7799 and similar schemes is to protect intellectual capital from an enormous array of threats, for the purpose of ensuring business continuity and minimizing risks while maximizing return on investments and opportunities.

The IT Security Functional Model

The deployment strategy for the IT security functional model is conceptually a *layered* approach. Because of the potential number of vulnerabilities that could create unauthorized pathways into a given enterprise network, implementing security measures and countermeasures at strategic network operating levels is the optimal defense strategy in thwarting the wily hacker. Figure 10–4 maps the proposed IT security functional model with a network example. The IT security functional model provides the blueprint that recommends an IT security architecture for practical deployment into your enterprise.

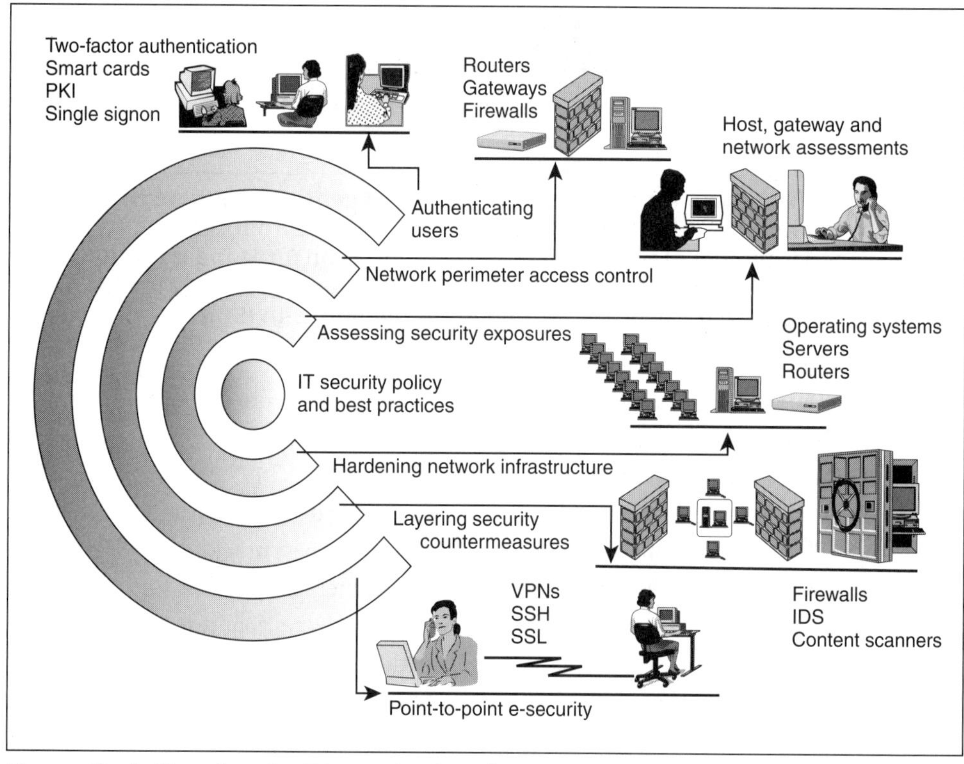

Figure 10–4 Mapping the IT security functional model

Hardening the Network Infrastructure for E-Security

Hardening the infrastructure is the most critical area to address in achieving effective network security. As with any architecture, doing so provides the necessary foundation for the security architecture to function effectively. Hardening the infrastructure means eliminating extraneous network services, default user accounts, and demos, as well as ensuring properly configured systems and network deployments.

Even with firewalls and perhaps even IDS systems in tow, some organizations are still susceptible to attack, because strategic network nodes or decision points, such as the routers or the operating system controlling online servers, are not sufficiently hardened, resulting in critical vulnerabilities being overlooked. (See Chapter 5.) Moreover, improperly configured/deployed security countermeasures are also susceptible to attacks. Enterprises, therefore, must exercise the necessary due diligence to ensure that the network's infrastructure, the resulting security architecture, and related controls are sufficiently hardened to provide the best foundation for delivering security measures holistically.

Controlling Network Access at the Perimeter

When you think of controlling network access on the perimeter, you immediately think of firewalls, and rightfully so. But the operative word here is *perimeter,* and this book has shown that although firewalls play an important role in the overall e-security scheme, the perimeter must be sufficiently fortified and robust to fulfill its important role in enabling e-business. In other words, firewalls play a key role in instilling the necessary trust and confidence that end users demand to operate safely on the Internet. This part of the e-security functional model answers the question, How effective is my firewall? and suggests a blueprint for deploying a firewall that will be resilient in a given computing environment.

Assessing Life-Cycle Security Exposures

In the IT components comprising the computing resources of your network, what vulnerabilities and exposures exist? How do you ever feel comfortable that the services, protocols, operating systems, and resulting applications are free from vulnerabilities and security threats? When are you most susceptible to vulnerabilities and exposures?

No matter how many security measures you deploy, if your network is laden with vulnerabilities, its security is precarious at best.

This area of the e-security functional model crafts a blueprint that addresses vulnerabilities and exposures in an ongoing basis. The resulting architecture that you deploy should be in response to the level of risks such vulnerabilities pose to your network and a tradeoff of cost versus acceptable risk. Without vulnerabilities, the hacker's ability to penetrate your network is greatly diminished, leaving less creative endeavors, such as social engineering and desk blotter raids, for IDs and passwords. In social engineering, would-be attackers pretend to be legitimate in-house IT support staff in order to confiscate user IDs, passwords, and other information to gain unlawful entry into a particular network. Most such tactics can be eliminated by instituting security policies that prohibit end users from providing critical login information by telephone or to an unfamiliar support person.

Layering Security Countermeasures

When information assets go online, the most effective protection entails deploying security countermeasures at critical *operating* layers within your network. One key layer of security is, of course, the firewalls on the network's perimeter. The firewall's security is like an electronic fence around your network. But what happens if the fence is scaled or tunneled under from below? How do you ensure that back doors do not materialize or breaches don't occur from inside the fence?

The blueprint from this part of the functional model focuses on strategic deployment of countermeasures on various operating levels *inside* the network's perimeter. The obvious choice for providing this type of security measure is an intrusion detection system. IDS systems can operate on the network or the host level. At the host level, IDS systems monitor end user activity to detect any business rules violations or unauthorized activity. Other security measures, such as e-mail scanners, can be implemented on the application level to monitor this user-demanding activity as well. Layered countermeasures allow you to protect all the strategic operating levels in your network, especially within the host and the application levels. The purpose of this part of the functional model is to protect your network inside its own doors.

Securing Point-to-Point Connections

This part of the functional model, which concentrates on protecting in-transit information/data could be the trickiest area to pin down. The type of security that you ultimately deploy depends on the parties involved in the communication and the type of information being transmitted.

For example, if you are building a Web server to support business-to-consumer transactions, you will most likely rely on an industry standard, such as the Secure Socket Layer (SSL) protocol, for providing a secure tunnel. The SSL protocol, supported by most popular browsers, encrypts the data for privacy, authenticates the connection, and ensures the integrity of the information. For business-to-business communications of remote, nomadic, or home users, a virtual private network could be the ticket. Or, for internal computing involving administrative activities, perhaps Secure Shell (SSH) or a VPN could be deployed for encrypted or authenticated administration. The bottom line is that if point-to-point communications must traverse untrusted networks or the potential for data compromise by unscrupulous insiders exists, in-transit security is warranted.

Authenticating the User Base

When networks are open, the perimeter nebulous, and concern is ongoing for unauthorized access to the system, making certain that the intended user base accesses the system could be the most daunting challenge for e-business security. Ensuring that users *are* who you think they are or should be is especially challenging if your network is accessible from many geographically dispersed locations.

The goal of this portion of the functional model is to map out login systems that provide the most effective authenticating mechanisms that could be reasonably cost-justified. Typically, an enterprise that has a user base of nomadic users who access the network from various geographic locations would probably have to spend more to ensure that those users are authentic than would an enterprise whose remote offices experience minimal staff changes.

If users must access multiple systems requiring the use of multiple logins, how do you ensure that users are authenticated under these circumstances? The security architecture that you deploy for authenticating the user base must take these issues under consideration. The architecture should also factor in the culture of the enterprise. If it has many administrative and clerical individuals, deploying a system that requires fancy login procedures may encounter resistance for wide-scale adoption. On the other hand, if the system is too simplistic for the culture, unauthorized users may slip into the network. Therefore, the architecture that you deploy must necessarily be effective, flexible, reliable, and resilient against the potential for compromise and unauthorized entry.

Deploying Effective E-Security Architecture: Hardening the Network's Infrastructure

The e-security functional model has been carefully laid out. In this section, we focus on building an e-security architecture, using the e-security blueprint as guidelines. The resulting security architecture, relating to each area of the functional model, incorporates IT best practices that optimize the effectiveness of the security measures themselves when initiated into service. Much of the best practices that provide the building blocks for the e-security architecture are reasonably sound, practical, or inherently cost-effective, enabling enterprises at various financial levels to deploy a security architecture that maximizes their networks' active defenses.

Chapter 5 provided an overview of the importance of eliminating unnecessary services, user and default accounts, and insecure native protocols and services from your network. Chapter 5 also talked about the importance of properly configuring key network components, such as servers/operating systems, gateways, and routers. This section explores best practices for hardening critical network infrastructure components. Specific guidelines are given for deactivating and disabling certain insecure services and protocols at the router through the access control list (ACL) list and/or the firewall. This section also discusses how to harden specific operating systems, such as Linux, UNIX, and NT, and critical servers, such as firewall gateways. (Appendix C includes recommendations for hardening Windows 2000 systems.)

Hardening Your Router

If you install a firewall and especially an IDS, you wouldn't necessarily have to be concerned about the kinds of sessions and related services allowed by router ACL lists, because the firewall and IDS system would provide all the security you need, right? Wrong. The types of services that are controlled by the firewall and related security measures should also be mirrored by the router's ACL list. The reason you have a firewall and especially an IDS is to prevent someone from getting in from the outside or opening a door from the inside, respectively. But what happens if the firewall and/or IDS is attacked by a DoS, for example? If the ACL list on the router allows access to certain insecure services, they could be exploited to allow an attacker to slip into the network while the firewall and/or IDS is busy fending off the DoS incursion.

Moreover, if you have a network with perhaps thousands of hosts, chances are that improper configurations, insecure services, default accounts, and other vulnerabilities may materialize on your network from time to time. These potential windows are opened *from the inside* to allow intruders in. Staying abreast of all the classes of vulnerabilities that may be introduced into your network is challenging, even with vulnerability assessment tools. Unfortunately, intruders have to find only one or a few unguarded windows/doors to gain unauthorized access into your network. If your router allows access to one or more of these vulnerabilities, the intruder would have to attack only the firewall and/or the IDS to slip into the network through the router. For these reasons, the router ACL list must be hardened to match the firewall and, perhaps, even the IDS to prevent intruders from slipping through unguarded windows and doors if you find your firewall and IDS come suddenly under attack.

Hardening your router requires knowledge and ability to work with ACL lists. Therefore, whether you are using a Cisco, Bay Networks, or other type of router, the services listed in Table 10–1 should be disabled or not permitted through external routers or those that are downstream from ISP's upstream routers.

Table 10–1 Router Security and Hardening Measures

Service/ Protocol	Description	Security Risk
SNMP	Verifies the operational status of routers	Creates security risks, so deactivate
UDP	A connectionless transport-layer protocol	Favored in many hacker exploits
TCP	The *connection*-oriented transport layer, in contrast to UDP, which verifies the connection	Favored in hacker exploits, but many key services, such as FTP, rely on TCP; if possible, deactivate
HTTP server	In Cisco routers, for example, allows remote administration	Can be modified through Web browsers and the right password
FTP	Usually enables hundreds of ports to accommodate various file transfer activities	If FTP required, deactivate unneeded ports

(continued)

Table 10–1 Router Security and Hardening Measures *(continued)*

Service/ Protocol	Description	Security Risk
TELNET	Creates a virtual terminal for connecting incompatible computers to the Internet	Exposes login and passwords in clear text as it traverses the network
Finger service	Determines whether active users are on a system	Allows attacks to occur without drawing attention to the system
BootP (Bootstrap Protocol) service	Enables users to discover their own IP addresses and those of servers connected to the LAN	A notoriously major security risk
ICMP	Used to handle errors and to exchange control messages	Used by several attack classes, so limit what ICMP messages are allowed
IP Unreachable message	Forces an ICMP type 3 error message to display when an ACL list drops a packet	Returning IP Unreachable error messages allow router fingerprinting or that access lists are used
IP redirects	Enables packets to be redirected from one router to another router to traverse and to exit networks	Enables hackers to engage in malicious routing to escape safety nets deployed in networks
IP directed broadcast	ICMP echo reply (normal ping) packets that are sent to a network's IP broadcast address	Smurf bandwidth attack
Source routing	The ability to specify the route packets can take from a source to a target host	Can establish a trust relationship between an attacker (source) and a trusted host

Table 10–1 Router Security and Hardening Measures ***(continued)***

Service/ Protocol	Description	Security Risk
Antispoofing	Rejects packets from external networks with source IP addresses belonging to your internal network	Packets with spoofed addresses signify a variety of hacker exploits
Private and reserved IP addresses*	Nonroutable IP addresses that are usually reserved for internal network activity and should not be accepted as source address from external networks	Masks security exploits when originating from external networks
Ingress filtering**	Denies spoofed IP packets by verifying that *source* IP addresses match the valid addresses assigned to the *source* network	Deters malicious insiders from launching attacks within the internal enterprise network
Password security	Encrypts password strings, especially for router configuration, to prevent viewing of passwords in clear text	Prevents confiscation of passwords by eavesdropping
User authentication	Provides various methods for authenticating administrators/users	Prevents unauthorized access to administrative-level control

*See Request for Comment (RFC) 1918 for list.

**See RFC 2267 for Ingress Filtering Review.

After certain critical functions have been disabled, other precautions, such as password security, should be instituted at the router as well. For example, during router configuration, encrypt password strings to guard against their being confiscated from system logs and related reports, such as electronic copies of your router configuration. Encryption service may be included with router utilities, so make good use of it to encrypt passwords. Note, however, that encryption provided through such services tends to be relatively weak, so remove encrypted password strings from written reports/copy.

User authentication for router administrators should also be carefully addressed. Depending on the size of the enterprise network, several router administrators may be required. Certain router manufacturers, such as Cisco Systems, allow administrative privileges to control the level of access to router functionality. These privileges can be set to allow various levels of control. Each administrator's associated login and password can in turn be authenticated, using MD5, for example, or other features provided by the manufacturer.

For direct and remote administrative management, other precautions can be taken as well. For direct access to the router through a console or an auxiliary port, use *autologout* functions for idle time. The administrator will be automatically logged off the router after a certain interval of idle time, usually 2–5 minutes. This prevents unauthorized access in case an administrator accidentally forgets to log out before leaving the router.

If SNMP is still preferred for remote management of the router to maintain the highest level of security in your network, consider out-of-band management, an IT technique that specifies setting up a separate network segment to provide security to accommodate critical network activities. To manage a router with SNMP, which is inherently insecure, consider deploying a three-interface router instead of one with two interfaces. The first two interfaces can be used to accommodate regular external and internal connectivity, respectively. Your access lists should disable SNMP to both external and internal hosts on the related interfaces. The third interface on the router should be set up to accommodate SNMP for remote management of internal router administration. By taking remote management of the router out of the normal band of network traffic, SNMP can still be used for router management on the enterprise level. Using SNMP through an out-of-band network segment provides the highest level of security and is the recommended practice for using SNMP safely.

Finally, synchronize your router with other hosts on your network through Network Time Protocol (NTP). Without synchronized time among the router, firewall servers, and switches, event correlation from log message timestamps is practically impossible. NTP allows its clients to authenticate related time sources, preventing attackers from spoofing NTP servers and ultimately manipulating the system clock. Having the time signatures in various hosts synchronized and authenticated through NTP will aid forensic investigation in the event of an attack.

Hardening Your Operating Systems

With optimum security architecture implemented in your router(s), the perimeter of your network's infrastructure is sufficiently fortified to take on attacks. In this section, we go behind the network's perimeter to harden other key components of the network's infrastructure: operating systems. Hardening the operating system is a critical phase in building the IT security architecture. Operating systems right out of the box usually include the extraneous services, default user accounts, related protocols, and utilities that possess the vulnerabilities and exposure that are susceptible to attack from untrustworthy network environments. Disabling and eliminating unneeded operating system services and utilities is the goal of hardening your OS and a recommended best practice in crafting resilient IT security architecture.

Linux, UNIX, and NT are the operating systems that are managing the millions of the Internet's host computers. Both Linux and UNIX are available in various flavors. There is only one NT, but various service pack levels can be in production in an organization at a given time.

Identifying what operating system is installed on your host(s) is one of the first steps a hacker will take to launch a successful attack against you. Identifying your operating system is called fingerprinting the stack. Tools are so adept at pinpointing an operating system that hackers can differentiate versions of the same operating system. For example, nmap can reliably distinguish among Solaris 2.4, 2.5, and 2.6.

Many security holes are dependent upon OS version. Further, if a hacker discovers that you are running Solaris version 2.51 or Linux 2.0.3.5 and if port 53 is open, you are likely running vulnerable versions of BIND. (BIND, or Berkeley Internet Name Domain, is the most widely implemented domain name service [DNS] on the Internet.) With this information, the hacker could enter your system with just a few minor modifications of code. However, if the operating system is hardened, your system will not be penetrated when the hacker tries to exploit known vulnerabilities associated with the operating system version your Internet hosts are running. For a review of remote detection of operating systems via TCP/IP fingerprinting, obtain the article of that name at www.insecure.org/nmap/nmap-fingerprinting-article.html.

Linux

Linux is fast becoming one of the most popular operating systems for e-business platforms, mainly because of its power and wide availability in the public domain. To harden the operating system, start at the beginning, with a clean installation.

Only with a clean installation can you ensure system integrity so that no unauthorized modifications or tampering with operating system components has occurred. Depending on the organization, this could prove to be a monumental undertaking if many operating system hosts are to be redone. If reinstalling every one of them is financially infeasible or beyond your level of resources, perhaps clean installations can be attained with the most critical production hosts. (For the record, don't forget to back up your data.)

Once Linux is reinstalled, never connect it directly to the Internet or to any other untrustworthy network, even when downloading operating system patches, updates, or upgrades. A PC should be designated for the expressed purpose of downloading such items. Similarly, a Linux-based production system should be attached to its own private network segment. When updating the Linux kernel, the PC with the OS downloads should in turn be connected to the isolated production network segment to update the OS. Handling your OS hosts and related downloads in this manner is a recommended best practice and a practical measure for fostering a secure computing environment.

Starting with the clean copy of Linux, the next step is to install the recommended security patches for your version of the operating system. Security patches, which are the *electronic inoculations* of OS security ills, are critical to fortifying an OS and should always be updated to maintain OS health. Without them, your systems can be easily compromised. Therefore, with your go-between PC configured for downloading and obtaining patches electronically over the Net, obtain the latest patches and reconnect to the private network where the clean OS host resides, and complete the security updates.

Two excellent sources for following bugs and system patches, especially for the Red Hat version of Linux, are Bugtraq@securityfocus.com and redhat-watch-list-request@redhat.com. Red Hat version 6.1 and higher includes an automated facility, up2date, for obtaining patches. This tool is highly customizable and easy to use, determines which Red Hat files requires updating, and automatically retrieves updates from the Red Hat Web site.

With the clean version of the OS fully installed and updated with security patches, the next steps involve the hardening activities. The four steps required to harden Linux are

- Disabling or eliminating unneeded or extraneous services
- Adding logging capability
- Fine-tuning certain files
- Installing TCP wrappers and IPChains

Disabling or Eliminating Unneeded or Extraneous Services

Linux is a powerful operating system that offers many useful services. However, many of them pose potential security risks to a given environment and consequently should be turned off. In addition to eliminating such services as Finger and BootP (see Table 10–1), also eliminate the services listed in Table 10–2.

You should be familiar with many of the services listed in Table 10–2. An emerging pattern suggests that these TCP/IP services in particular are inherently insecure whether they are executing at the router or the operating system level. In addition to the R services[3] you may be familiar with, Linux—in particular, Red Hat 6.0—possesses some other R services that aren't so commonly known. Rusersd, rwhod, and rwalld should also be eliminated. In general, try to avoid running any R services altogether, as they pose a serious security risk by providing intruders with considerable latitude for entering your network illegally from remote locations.

Other Linux scripts you should eliminate unless they are absolutely necessary include dhcpd, atd, pcmcia, routed, lpd, mars-nwe, ypbind, xfs, and gated. (For an explanation of these services, refer to Red Hat 6.0 documentation or Webopedia: www.webopedia.com, the online computer dictionary for Internet terms and computer support.

The preceding scripts and services and those listed in Table 10–2 are installed by default, initialize when the system is booted, but are not critical to system functioning. The bottom line: If you don't need them, turn them off. If you do need them, make certain that you exercise the proper controls through your security policies and firewall rule sets to monitor and/or limit access.

Logging and Tweaking

With as many services as possible eliminated, you are well on your way to achieving a hardened operating system kernel, the most important module of the operating system. The fewer the services the kernel has to manage and the fewer the processes and/or tasks, whether legal or illegal, to be supported, the more efficiently your operating system will perform.

Linux has excellent logging and supports running processes—executing programs/scripts—well except for FTP. If you must use FTP, lock down all related

3. R services, short for RPC (remote procedure calls) services, are subroutines that allow programs on one computer to execute programs on a second computer. They are used to access services connected to shared files.

Table 10–2 Services/Scripts to Be Eliminated/Deactivated in Linux

Service/ Script	Description	Security Risk
Post Office Protocol (POPD)	Used to store incoming e-mail on end users' computers	History of security issues; if available, hackers use to launch e-mail attacks against users
Internet Message Access Protocol (IMAPD)	Used to store incoming e-mail messages on a central mail server	History of security issues; eliminate
RSH (remote shell)	Enables remote execution of printer commands from external IP addresses	Implies a trusted relationship between hosts; major security risk if accessed by external users from spoofed addresses
APMd (Advanced Power Managment daemon)	Used only for laptops	Vulnerable script, so delete
XNTPd (X Network Time Protocol daemon)	Network time protocol	Can be used to alter the system clock
Portmap	Required if you have any RPC services	Eliminate RPC and dependent files; RPC services alter file permissions and steal password files, remotely
Sound	Saves sound card settings	Vulnerable script; delete
NETFS	The NFS client, which mounts file systems from an NFS server	Vulnerable script; delete

Table 10–2 Services/Scripts to Be Eliminated/Deactivated in Linux *(continued)*

Service/ Script	Description	Security Risk
Rstatd	Communicates state changes between NFS clients and servers	Enables an intruder to spoof a legitimate rpc.statd process; R services are too accommodating to remote users, so delete them
YPPasswdd	Necessary for NIS servers	An extremely vulnerable service
YPserv	Necessary for NIS servers	An extremely vulnerable service
SNMPd	Provides operational status and detailed information on network components	SNMP used to compile operational and detailed information on your system
named	Used to set up the DNS server	May contain a vulnerable version of BIND; upgrade to secure version, if needed
NFS	Allows mapping of remote directories as extensions to local user files	Allows intruders to browse entire file systems for poorly secured directories
AMD (automountd)	Allows mounting of remote file systems	Used with statd to gain administrative privileges in exploited host
Gated	Used to run third-party routing protocols	A vulnerable script; delete it
Sendmail	Widely used to implement e-mail in TCP/IP networks	A history of security problems; delete if not being used for e-mail
HTTPd	The Apache Web server daemon	If used, implement the latest version
INNd	The network news daemon	Allows hackers to attack systems through network news service protocol
linuxconf	Enables a user to configure Linux through a standard browser	Extremely vulnerable

services on the FTP server. To ensure that Linux logs FTP activity properly, edit inetd.conf, the configuration file that contains all the services allowed to run in your network. A hardened Linux host has minimal network services and/or programs enabled. When properly configured, FTP logs all FTP sessions and user commands to syslog, the main host logging function.

Other logging options are available for FTP as well. If other logging activity is desired, make certain that they are properly secured. For example, xferlog, which records all FTP uploads and downloads, is a great source for determining what intruder tools may have been installed or information downloaded if your system is compromised. Logs, especially syslog, can record user passwords under certain circumstances, allowing them to be confiscated if the correct safeguards aren't deployed.

An important area that should be tweaked is file administration, and securing the password file is one of the most critical tasks in providing an optimal level of security. In Linux Red Hat version 6.0 and in UNIX, the /etc/passwd file must be fortified. This file is a database that stores user accounts and their associated passwords. In Red Hat version 6.0 and later, user passwords are stored as hashes and securely placed in a file that is accessible only from root, accomplished by using the /etc/shadow utility. You can use either the default hash crypt (3) or message digest (MD) 5.

MD5 provides an even greater level of protection. By storing user passwords in their hash values, etc/shadow protects passwords from being easily accessed and worse, cracked. Red Hat version 6.0 automatically converts the password file into "shadow" passwords by default. For other versions of Linux, only a simple command sequence is required to convert user passwords to their hash values and stored in the etc/shadow file. Like clockwork, one of the first actions a hacker will take after gaining a root compromise is to access the etc/passwd file to confiscate user passwords and log-on IDs. Therefore, this is one of the most important actions to take in securing your Linux host.

You are not done tweaking the etc/passwd account yet. This file contains default user accounts for news services and FTP, or anonymous FTP. If you are not planning to run your hardened Linux host as a newsgroup server, which requires NNTP (Network News Transport Protocol), remove the news user. Be sure that related files, such as etc/cron.hourly, are updated, because this file looks for user "news." Also remove the user account FTP, which allows anyone to access an FTP server as a user called anonymous, or FTP. An anonymous FTP user can choose any password—typically, the client's host name.

Other files you should tweak after the /etc/password file are /etc/ftpusers, /etc/securetty, and /etc/issue. The /etc/ftpusers file function, contrary to what you might guess, does not enable FTP users but instead *prevents* users—especially system users, such as root or bin—from attempting FTP sessions. Linux provides this file by default. As a rule of thumb, you never want root to possess the FTP ability to the system. Consequently, if you do *not* want users or accounts with the ability to initiate FTP sessions, include them in the /etc/ftpusers file.

You should also ensure that root cannot TELNET to the system. This forces users to log in as themselves and to use other authorized means to access root. The file /etc/securetty lists contain all the virtual terminals ttys root can connect to. Such listings as tty1, tty2, and so on, restrict root logins to local access. Such listings as ttyp1 allow root to log in to the system remotely. To restrict what root can TELNET to, make certain that this file is modified accordingly and the necessary diligence exercised for ongoing control.

Finally, the /etc/issue file is an ASCII text banner that greets users who log in by TELNET. This file is typically used to display legal warnings whenever someone attempts to log in through TELNET to your system. By default, however, Linux creates a new /etc/issue file on every system reboot. To use the same legal warning for every reboot, Linux allows you to modify the . . ./init.d/S99local file.

Using TCP Wrappers and IP Chains

As a manager of IT security, one of your key responsibilities is to manage your server once it goes online. If the server must be managed remotely, your connection should also be secured. Despite expending considerable effort to harden your operating system on production units, leaving an unsecured connection could pose a security risk to the network. Two options are recommended for establishing a *secure* remote management connection into your Linux server: TCP wrappers and a secure tunnel, or Secure Shell (SSH).

A TCP wrapper is a binary, or executable, file that wraps itself around inetd-controlled services, such as HTTP, FTP, or SMTP. Recall that inetd is a powerful service that ensures that executing processes receive system support in terms of protocols and related services. Inetd listens on Linux ports for the requested services and makes them available to the connection in accordance with the inetd.conf. file. For Linux users, TCP wrappers are available by default on installation.

Access control through TCP wrappers is achieved in conjunction with two files: /etc/hosts.allow and /etc/hosts.deny. When a user attempts to connect to the Linux server, Linux launches the wrapper for inetd connections. The system enables the wrapper to verify the connection attempt against the access control lists in the two files. If the connection is permitted, TCP wrappers relay the connection to the appropriate executable, such as HTTP or TELNET, and the connection ensues. If the connection parameters reside in the /etc/hosts.deny file, the connection is dropped.

TCP wrappers log all attempts and deploy secure programs and utilities. Connections to services operating with TCP wrappers are logged to the Secure log file.

With all its power, however, TCP wrappers will not protect your network from sniffing and do not provide encrypted connections. Intruders can still capture restrictive information, including passwords transmitting in clear text over the network. If you feel that your network is sufficiently private and fortified against outside intrusions, TCP wrappers would be adequate in a trusted computing domain. However, to guarantee a higher level of privacy for remote administration, Secure Shell (SSH) may be more appropriate.

Typically, TELNET, FTP, or Rlogin is used for remote management activities. These services have a history of security problems. First, passwords are transmitted across the Internet in clear text. Therefore, SSH is the preferred method for managing your Linux server from a remote location. In particular, SSH is growing in popularity becuase of its ability to provide a variety of encrypted tunneling options, authentication methods, and compatibility with popular operating systems.[4] Like TCP wrappers, SSH also provides logging capability. SSH encrypts all remote management traffic, including administrator passwords, after authenticating the user. With SSH providing the secure connection, you can ensure that administrators are authentic, thwart eavesdropping, and eliminate connection hijacking and related network attacks.

Finally, IPChains is to Linux hosts what ACL lists are to routers. The IPChains utility is packet-filtering software that is included in the installation kit in Linux Red Hat version 6.0 and later. IPChains software is similar to Cisco ACL lists in form and functions much like a firewall. The software controls what packets can come in and out of your Linux box in a network and stand-alone environment.

Implementing the precautions covered in this section will give you a strong base level of security. The key to having a secure, or hardened, Linux system is making sure that the minimal software in terms of services is installed, default

---

4. OpenSSH is the most popular SSH version. Use OpenSSH to replace rlogin, TELNET, and FTP for remote management operations. The product has been enhanced, and many of the problems with earlier SSH have been corrected. For more information on OpenSSH, go to http://www.openssh.com.

user accounts eliminated, and security architecture deployed in layers, using TCP wrappers or SSH, IPChains, and hashed password protection. (To *automatically* secure Linux hosts, check out Bastille Linux, a PERL script that provides step-by-step instructions.)

Additional steps, such as password protecting the BIOS (basic input/output system) and the system to restrict physical access, may be warranted, depending on the challenges you may face in your computing environment.[5] In the final analysis, no matter what you do, no system can be made 100 percent secure. However, deploying the security measures and architecture discussed here are best practices for considerably reducing critical security risks to your e-business computing platforms.

UNIX

The process for hardening UNIX is the same as that for hardening Linux.[6] Although Linux is a more feature-laden derivation of UNIX, the recommendations about eliminating services, tweaking password files, and deploying security architecture in layers generally mirror those recommended for Linux. Linux Red Hat version 6.0 and higher provides certain default services, such as the Shadow password utility, use of MD5 hashes automatically with Shadow, and TCP wrappers. Your particular version of UNIX may provide such services by default.

As with Linux, hardening a UNIX host begins with a clean install of UNIX. If your IT environment consists of many UNIX hosts, reinstalling clean copies of your version of the UNIX operating system may be too costly. However, to minimize security risks stemming from operating system vulnerabilities, find an acceptable tradeoff that perhaps would entail hardening only mission-critical hosts. A limited effort may stand a better chance of being approved by executive management.

Starting with a clean install of your operating system eliminates any doubts about what could be or might have been exploited. With UNIX, the kernel, binaries, data files, running processes, and system memory all have associated vulnerabilities. Therefore, to ensure that the operating system is free of exploits, it should be reinstalled from the distribution media provided by the manufacturer of your version of UNIX.

After the operating system has been reinstalled, the next step is to apply the full suite of security patches for your operating system. One of your most critical ongoing tasks is staying on top of all the security alerts for your operating system.

---

5. For an excellent overview of securing Linux hosts, see the article "Armoring Linux" by Lance Spitzner at www.enteract.com/~1spitz/papers.html.

6. For more information on hardening your UNIX system, see "UNIX Configuration Guidelines" at www.cert.org/tech_tips/UNIX_configuration_guidelines.html.

To accomplish this, make it a practice to regularly visit the Web site of the manufacturer of your version of UNIX. As recommended in the Linux discussion, never connect your host with the clean copy of the operating system directly to the Internet or any untrustworthy network to download security patches. A go-between PC with the necessary security precautions—virus software, personal firewall, and so on—should be used for this purpose.

With the security patches installed, connect your go-between PC to the clean operating system hosts, which should be connected to their own separate network segment. With your operating systems updated, you can begin the process of hardening your UNIX hosts, which consists of

- Disabling or eliminating unnecessary services
- Enabling and securing logs
- Tweaking certain files
- Adding TCP wrappers and security in layers

Disabling or Eliminating Unnecessary Services

The same services identified as unsafe under Linux are unsafe under UNIX and so should be disabled. Table 10–3 lists all the insecure services that should be disabled

Table 10–3 UNIX Services to Be Disabled/Deactivated

Services	Description	Security Risk
TELNET	Used to access various ports for various system processes	Replace with OpenSSH.
FTP	Popular for managing file servers for distributing documents	Deactivate it if you don't need it, or replace with OpenSSH.
Finger	Determines whether an active system is unattended	Allows hackers to gather information for later clandestine exploits; should generally not be run.
Bootstrap Protocol (BootP)	Allows users to determine their own IP addresses and those of other hosts attached to a network	Deactivate if not needed.

Table 10–3 UNIX Services to Be Disabled/Deactivated ***(continued)***

Services	Description	Security Risk
Comsat	An alternative service to POP and IMAP, notifies users of incoming mail via Biff	If other methods are being used to retrieve mail, disable.
Exec	Allows remote users to execute commands without logging in	Deactivate.
Login	Enables remote users to use rlogin and, if supported by an rhosts file, to do so without a password	For remote connections, use SSH and TCP wrappers instead.
Shell	Allows user to execute r commands remotely	If the function is required, disable and use TCP wrappers.
Netstat	Provides network status information to remote hosts	Can be used locally on the system, but disable for remote access.
Systat	Provides system status information	Disable.
Talk and Ntalk	Enables communications between local users and remote users on other systems	Disable.
TFTP (Trivial File Transfer Protocol)	A miniversion of FTP	If required, disable and use TCP Wrappers.
Time	Provide synchronized timestamps, which are critical for forensic investigation	Use XNTP instead.
UUCP (UNIX-to-UNIX copy program)	Used to transfer files from one UNIX system to another	Disable.

or eliminated from indetd.conf and related files. The table describes only some of the most exploited UNIX services. To stay on top of other UNIX services and utilities that may be exploited, check the Web site of your particular UNIX manufacturer, along with such security sites as CERT/CC and SANS Institute, on a regular basis. Another source for staying abreast of UNIX vulnerabilities is the *UNIX Insider* articles.[7]

Enabling and Securing Logs

UNIX systems offer options for logging user sessions. Perusing logs on a regular basis could prove to be a mundane, tedious activity. However, it is a necessary evil and perhaps even a best practice because logs show the patterns of use for your network. Familiarity with them will help you discern abnormal or suspicious patterns of use, which often prove to be the smoke signals of potential security fires. Following are common UNIX log file names and descriptions.

- *Syslog,* the main logging facility for UNIX systems, allows an administrator to log session activity in a separate partition or host. Under certain circumstances, it may be necessary to look in the /etc/syslog.conf file to determine where syslog is logging messages.
- *Messages,* the log designed to capture session information. In the event of an incident or intrusion, this log would be instrumental in revealing anomalies in and around the suspected time of the incident or intrusion.
- *Xferlog,* used to log data from FTP sessions. If FTP must be used, xferlog may assist you in determining when any uploads or downloads, for example, of intruder tools or unauthorized information, respectively, have occurred.
- *Utmp,* file that records binary information on every currently logged-in user. One way of accessing this data is through the who command.
- *Wtmp,* the file modified whenever a user logs in or out or when a machine reboots. Gleaning useful information from the binary file requires a tool such as last, which creates a table that associates user names with login times and the host name that originated the connection. Output from this tool helps you discover unauthorized connections, hosts that are involved, and user accounts that may have been compromised.

---

7. A comprehensive listing of articles by subject can be accessed at www.UNIXinsider.com/common/swol-siteindex.html. Back issues of *UNIX Insider* can be explored at www.UNIXinsider.com/common/swol-backissues.html. Home page is www.ITWorld.com/comp/2378/UNIXInsider/.

- *Secure,* the file to which some UNIX versions log TCP wrapper messages. Specifically, whenever a service that runs out of inetd with TCP wrappers supports a connection, a log message is appended to this file. This log reveals connections from unfamiliar hosts, especially for services that may not be commonly used.

An important best practice is securing your system logs. Logs are the electronic reconnaissance files of your network. However, your logs are not useful if you can't trust their integrity.

One of an intruder's first items of business is to alter your log files to cover his or her tracks or, worse still, to control logging by installing a Trojan horse. Hackers use a rootkit, such as Cloak, to wipe out their tracks recorded in system logs. Regardless of how secure your system may be, logs cannot be trusted on a compromised system. More important, some intrusions might erase your hard disk. For these reasons, a dedicated server that captures logs remotely from other key systems should handle logging activity.

A dedicated logging server can be easily built inexpensively under Linux. Make sure that you turn off all services and follow the related recommendations for hardening Linux. Access should be through the console only. Therefore, block port 514 UDP, which is used for logging connections at the firewall that controls your Internet connection. This prevents your dedicated log server from receiving bogus or unauthorized logging information from the Internet. Next, recompile syslogd to read a different configuration file, such as /var/tmp/.conf. To accomplish this, change the source code in /etc/syslog.conf to the new configuration file, which in turn is set up to log both locally and to the remote log server. As a decoy, the standard copy of the configuration file should, on the system in question, point to all local logging. This process should be repeated for all systems that you want to have logging into the dedicated, remote server. Once in place, *local* system log files can be regularly compared against the *remote* log files to monitor whether local logs are altered.[8]

To assist in managing the mounds of data that can be generated from logging activity, check out a logging tool called Swatch, or the *Simple Watcher* program, a device used for monitoring and filtering UNIX log files. Swatch watches

---

8. For a complete discussion on this topic, refer to the article "Know Your Enemy: II" at www.linuxnewbie.org/nhf/intel/security/enemy2.html.

for suspicious data in response to set parameters; when it encounters these patterns, the program can take certain actions to alert security personnel.[9]

Tweaking Certain Files and Services

One of your most important activities is protecting your password file against directed attacks. As pointed out in the Linux discussion, the /etc/passwd file stores user accounts and associated passwords. Passwords may be stored in DES encryption format. You would think that encrypted passwords would be safe, even if the file is somehow compromised or stolen. Unfortunately, however, that is not the case. A hacker who could somehow steal your password file would promptly move or copy it to another machine to run a brute-force dictionary search attack, using a program such as Crack against the password file. (Crack is a freeware tool that should be used regularly by system administrators as a password-auditing tool to ensure that users are using *uncrackable* passwords.[10]

Research has shown that these tools can crack 20 percent of, for example, DES-encrypted passwords of a user base of a medium/large enterprise in approximately 10 minutes. The Shadow password utility, on the other hand, stores a hash of the passwords in a separate file that is not world-readable. Shadow may be available in the distribution media of your version of UNIX. Consult your documentation, or call your UNIX manufacturer to find out if it is provided by default or from a distribution Web site.

Sendmail can be exploited to attack the /etc/passwd file. Sendmail's problems are well documented and have plagued the Internet for some time. Certain Sendmail vulnerabilities can be exploited to steal a copy of the UNIX password file.[11] Sendmail, version 8.7.5 fixes many of these vulnerabilities. However, new Sendmail exploits surface in the wild from time to time, although older versions of Sendmail, which have not been modified with the latest security patches, receive the majority of attacks.

All versions of Sendmail—whether or not updated with security patches—should be deployed with *Sendmail restricted shell (smrsh)*.[12] In addition to blocking

---

9. For more information on Swatch or to obtain a freeware copy, go to ftp://ftp.stanford.edu/general/security-tools/swatch/.

10. Crack is available from ftp://coast.cs.purdue.edu/pub/tools/unix/pwdutils/crack or ftp://info.cert.org/pub/tools/crack.

11. For a review of Sendmail vulnerabilities, go to www.cert.org/tech_tips/passwd_file_protection.html.

12. Smrsh can be obtained at etiher ftp://info.cert.org/pub/tools/smrsh/ or www.sendmail.org.

attacks on the UNIX password file, smrsh can help protect against another well-known Sendmail vulnerability, which allows unauthorized remote and local users from executing programs as any system user other than root.

The UNIX password file can also be comprised by TFTP (Trivial File Transfer Protocol), which confiscates the file. Safer alternatives are SSH and even a *wrapped* FTP. Because of its simple application—its ability to easily exchange files between separate networks and no security features—deploy TFTP, if required, with restricted access provided by TCP wrappers.[13]

Some UNIX distributions, such as IRIX from Silicon Graphics, distribute default system accounts without passwords assigned. After installations, these accounts are often forgotten, creating a vulnerability. The accounts in question include IP, demos, guest, nuucp, root, tour, tutor, and 4Dgifts. If you plan to use any of these accounts for network access, assign them strong passwords immediately.

A favorite tactic of intruders is to exploit system default passwords that were not changed after installation, such as accounts with vendor-supplied default passwords. Thus, make certain that all accounts with default passwords assigned on networking equipment and computer systems are changed before deploying them. Also, software updates can change passwords without creating any attention to that fact, so be mindful of this and check passwords after product upgrades.

As a rule of thumb, a single password should not be used to protect multiple accounts for a given user. If an intruder using a packet sniffer, for example, is able to confiscate a shared password that has been transmitted in clear text, all accounts sharing that password are compromised. Ensure that each account has its own unique password. This may require deploying hand-held or software tokens that generate one-time passwords. Usually, these devices accompany strong authentication methods that are suited for authenticating users across untrusted networks to access sensitive enterprise resources, such as intellectual capital, name servers, or routers.

Tweaking Other Critical Files

Some UNIX manufacturers' default state is to trust—allow user access—from other systems. These vendors issue the /etc/hosts.equiv files with a plus sign (+) entry. Remove the plus sign from this file; otherwise, your system will trust all other systems by allowing access to any client host. Similarly, the .rhosts file may

---

13. If your version of UNIX does not ship with TCP wrappers or is not available from the manufacturer, obtain the program from ftp://ftp.porcupine.org/pub/security/ or www.larc.nasa.gov/ICE/software-list/descriptions/tcp_wrapper-7.2.html.

be supplied with a plus sign entry that should be removed. This file is used to list host/user pairs that are permitted to log in and by rlogin and rsh, making .rhosts insecure by association. Both files should never be world-writable.

Many UNIX versions come preconfigured to allow secure, root login access through any TTY device. Depending on your release, check either the /etc/ttys or the /etc/ttytab file. The *only* terminal that should be set to secure is the console. Disallow secure logins from everything else.

If you must use FTP, make certain that anonymous FTP is configured correctly. Follow the instructions provided with your operating system manual to properly configure file and directory permissions, ownership, and group. Note that you should never use your standard password file, /etc/passwd, or group file for this service. Moreover, the anonymous FTP root directory and its two related subdirectories, etc and bin, should never be owned by FTP.

In general, ensure that you regularly check the protections and ownership of your system directories and files. Whenever new software is installed or verification utilities are run, these procedures can cause file and directory protections to change. Such changes create vulnerabilities. Thus, after such procedures are run, make certain that file and directory protections are set according to the recommendations provided by your system's documentation. Use a freeware package called COPS (Computer Oracle and Password System) to regularly check for incorrect permissions on file directories.[14] COPS is a collection of programs that provide a variety of helpful services, including host vulnerability assessments.

Adding TCP Wrappers and Security in Layers

As with Linux, UNIX system administrators have the option of providing security in layers throughout the UNIX computing environment. Once the operating system is hardened on as many operating units as possible, additional steps should be taken to institute other critical security layers. For remote administration, for example, TELNET, FTP, or Rlogin should be used with TCP wrappers if transmissions must traverse regular operating domains for remote administration. (Recall that TCP wrappers are similar to access control lists (ACL) for routers and provide the same functionality.) Check your documentation and/or software distribution media to see whether TCP wrappers were provided in conjunction with the regular OS modules. Alternatively, if remote administration will traverse untrusted networks, perhaps OpenSSH, a robust version of SSH, can be used for attaining

14. COPS can be obtained from ftp://coast.cs.purdue.edu/pub/tools/UNIX/scanners/cops/.

privacy for remote administration activity. OpenSSH provides authentication and encrypted tunneling for all transmissions, including passwords.

With TCP wrappers and SSH used for remote administration, an effective level of security can be achieved. Strong passwords and permissions protect systems if administration is conducted directly at the console. User access is controlled by solutions providing strong authentication and "tweaking" to restrict terminal access except at the console. Extraneous or unused services are eliminated or minimized to control the incidence of vulnerabilities while systems are in operation or online. With security measures implemented in layers, as summarized here, you institute some of security's best practices to operate safely in e-business environments. Figure 10–5 summarizes a layered security architecture approach for UNIX nad Linux systems.

Finally, as an extra recommended precaution, consider security tools, such as Tripwire, to provide integrity checking for critical programs, such as UNIX

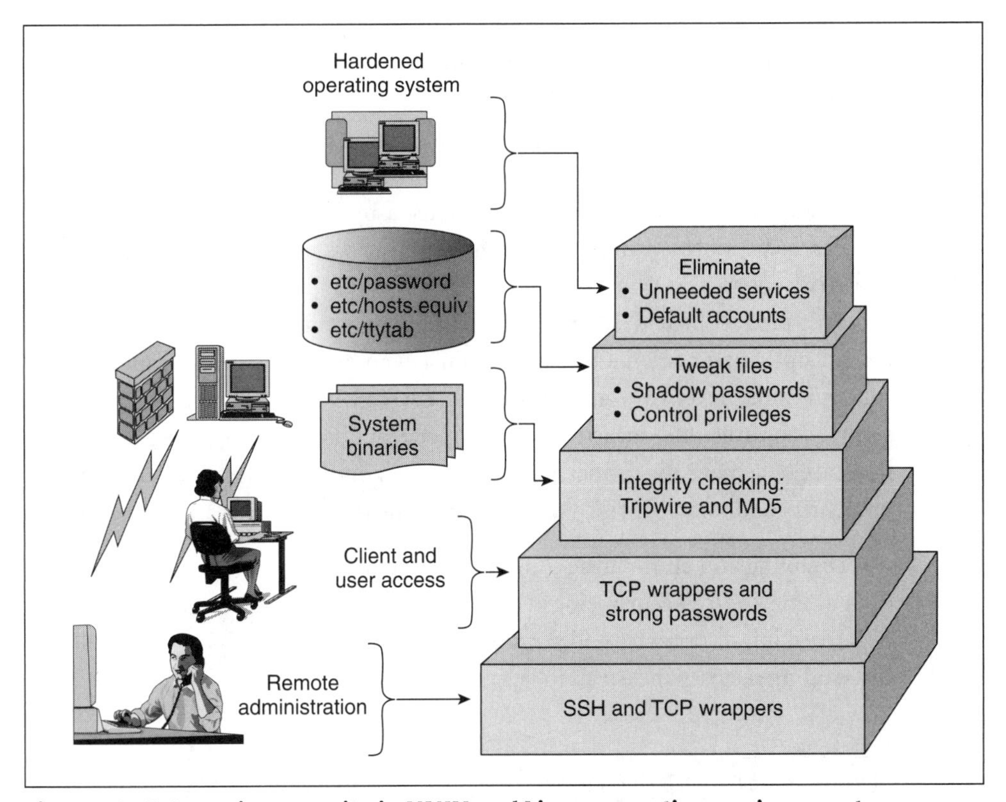

Figure 10–5 Layering security in UNIX and Linux operating environments

binaries.[15] Some hacker exploits modify system binaries and/or replace them with Trojan horses. Tripwire builds a database of files and directories you would like to monitor or, at a minimum, the programs and binaries that are typically replaced after a system compromise. (See the section Analyze the Intrusion in Chapter 8.) Each time Tripwire is run, which should be regularly, it flags deletions, replacements, and additions and creates the related log file. Tripwire builds the list after comparing its findings to the database of files and directories created when it is deployed. Tripwire enables you to immediately spot changes in your key operating files to take the appropriate actions, up to and including reinstalling the compromised host to ensure its integrity.

An alternative to Tripwire is MD5 checksums.[16] The MD5 checksum algorithm, designed by RSA Data Security, can be used to take a hash value of each of your system files, especially system binaries. The MD5 hash of a given file is that file's unique fingerprint. The manufacturer of your version of UNIX may be able to give you the list of checksums for your system files, or you can create the list yourself. As with Tripwire, MD5 checksums should be run on a regular basis to track any changes to your critical system files. MD5 is a freeware utility that is assigned standards track RFC 1321.

NT

Windows NT (New Technology), widely used by businesses worldwide, is one of Microsoft's most advanced *Windows* operating systems. (See Appendix C for a related discussion on Windows 2000.)

This section provides some general guidelines and best practices to secure the NT operating system. Of course, security can be instituted on a much wider, more granular level over and above that covered here. Nevertheless, some NT security controls and measures may impede operational processes. Therefore, take care to achieve the proper tradeoff between implementing a reasonable level of security and tolerable risk. Again, security measures can be categorized as

- Disabling and eliminating unneeded services
- Adding and/or securing logs
- Tweaking certain files and services
- Adding security measures in layers

---

15. Tripwire can be obtained from ftp://coast.cs.purdue.edu/pub/tools/UNIX/ids/tripwire/.

16. For more information on MD5 checksums or to obtain a copy, refer to the CERT advisory on www.cert.org/advisories/CA-1994-05.html.

Disabling and Eliminating Unneeded Services

Windows NT services are programs that the system uses on start-up. As with all operating systems, NT services typically run in the background, servicing requests from users and the network simultaneously. By default, a large number of services are available through NT after installation. These services are typically installed to and used by the omnipotent NT system account, which is accessible from the NT Services Control panel.

At this point, a distinction should be made between eliminating NT, or *operating system,* services and *network* services provided by native network protocols, such as TCP/IP and NetBEUI (Network BIOS Extended User Interface). On installation, NT makes available both operating system services, such as the alerter, the computer browser, the directory replicator, and so on, and typical network services provided through, for example, the TCP/IP stack, which runs native under NT.

By default, many NT, or operating system, services possess security-sensitive rights, any one of which could subvert overall system security. Also, many of the network services are susceptible to packet-level attacks. Thus, take care to eliminate as many *unnecessary* NT services as possible, especially those that are likely to be attacked while traversing untrusted networking environments. To be sure of the services that you don't need, verify their function in the NT system documentation before disabling them. If any of the following services are unnecessary, disable them:

- Alerter
- Computer browser
- DHCP (dynamic host configuration protocol) client
- Directory replicator
- FTP publishing service
- Internet information service admin service
- Messenger
- NetBIOS interface
- Net log-on
- Network DDE (dynamic data exchange)
- Network DDE NSDM
- Remote procedure call locator
- Remote procedure call configuration
- Server
- Spooler (not required if a printer isn't directly connected)

- TCP/IP NetBIOS helper
- WINS (Windows Internet naming service)
- Workstation
- World Wide Web publishing service

The IP services that are eliminated/disabled under Linux/UNIX should also be eliminated under NT. TELNET, FTP, BootP, R services, and their associated ports should be disabled to prevent access from untrusted networks. In fact, NT enables network managers to configure the hosts to mirror the ports that are enabled at the firewall. Because TCP and UDP services can be accessed and controlled through the Control Panel menu, other security measures, such as allowing only certain administrators to control network services, limiting access rights of all other users to perhaps read-only. Moreover, using strong administrator passwords for general access and for screensavers works to provide holistic security precautions.

Enabling and Securing Logs

NT provides an efficient logging system that can capture a significant number of activities in its security logs. NT's auditing system gives security managers full control of what type of information can be audited and how much of it to log. When setting up your logging function, you should consider the amount of time and resources administrators can devote to the analysis and maintenance of security logs. NT supplies only a simple viewer for analysis of its logs. Thus, for in-depth log analysis, NT fully accommodates third-party analysis tools. As a best practice, the security log should be supported by a dedicated hard disk partition, with access restricted to authorized security personnel.

The NT logging capability allows the auditing of

- Log-on and log-off activity
- Start-up, shut-down, and system events
- Specific files, such as executables or dynamic linking libraries
- User and group management

NT also allows auditing of object access, process tracking, and use of user rights.

However, deriving meaningful patterns of any suspicious behavior requires analysis tools. Make it a practice to review logs periodically for unexpected or suspicious activity. If necessary, invest in the analysis tools that you may need to glean useful information from the logs and ultimately a return on your investment.

One final point about NT logging: Guest and unauthenticated users can access system and application logs—not the *security* log—by default. Although they possess read-only rights, these users can nonetheless discern useful information about the operating platform to subvert security. To disable this feature requires a simple modification to the application and system event log files in the NT registry, a database that *Windows*-type operating systems use to store configuration information. User preferences, display and printer settings, hardware, and operating system and installed applications parameters are typically written to the registry on installation and setup activities. When editing the NT registry, usually with regedit.exe, take great care that modifications are implemented correctly, because errors to the registry could disable the computer.

Tweaking Certain Files and Services

On installation, NT makes available many services in the system root directory: WinNT. Although Microsoft has instituted reasonable security precautions, trade-offs against optimal security were instituted to maximize compatibility with potential applications out of the box. However, to maximize security for your NT system, you should tweak certain files, especially WinNT, which has inherent access control mechanisms that, oddly enough, are called access control lists. NT's ACLs are similar to TCP wrappers on UNIX systems, but the application of ACLs provides more granular control over user privileges, and hence access, rights than TCP wrappers do.

In general, authenticated users should be granted read-only rights when accessing system files. After a standard NT install, the Public users group—which usually means *everyone*—can access certain files in the WinNT system root directory and registry. *Everyone* might be both unauthenticated and authenticated users or a subset of authenticated users. Whichever way you define this group, it should be granted only *read-only* privileges, because NT also provides *write* privileges to this group by default after a standard install. Public exercises read- and write-access privileges to the WinNT directory during an install of new software or hardware or for maintenance.

For obvious reasons, you do not want *everyone* to have access to WinNT directory trees and the registry. However, you also do not want to lose the capability to install applications on the local level, especially in large networks with multiple domains. Because NT allows you to create user groups, a trusted application installer, or App installer group should be created for the purpose of installing and maintaining software and hardware. This group would not carry the full privileges of an administrator, but at least you retain a potentially critical capability on the local level while limiting a perpetrator's ability for a root compromise of the

WinNT system directory and the registry. Finally, review file and directory permissions on C:\ and C:\Temp, and assign the permissions on these directory trees recommended by Microsoft.

When formatting storage volumes, use NT's NTFS (NT File System) format instead of the more common FAT (file allocation table) format. NTFS supports a feature called *spanning volumes,* which enables files and directories to be spread across several physical disk drives. Unlike FAT, NTFS incorporates the use of ACLs, providing a secure storage solution, and allows permissions to be set for directories or individual files. The FAT system, on the other hand, possesses an inherent vulnerability that allows anyone to reset FAT files with a read-only permission. Lacking a compelling reason to use FAT—and there seldom is on NT—format all potentially capable volumes with NTFS.

As with any operating system, institute a strong, effective password policy. NT has been widely criticized for network password exposure. Although it stores and uses the *hash value* of user passwords, NT is still susceptible to brute-force attacks from the confines of unfriendly local and remote networking environments. Furthermore, certain commands, such as RDIDSK, provide weaker protection, which leads to compromising security of stored hash values. NT usually stores the hash value of user passwords, in the user account database of the NT registry, also known as the SAM.

To strengthen the protection of locally stored hash passwords, use the NT's native SYSKEY command to configure the system so that user password hash values are encrypted with a 128-bit encryption algorithm for extra protection. Note that, according to many reports in the wild, NT's hashing algorithm is rather widely known, making the use of SYSKEY all the more important.

Devising an enterprisewide password policy and practice is challenging enough, especially in a diverse user population. NT provides several mechanisms to ensure that users are adhering to acceptable password practices. The three areas subject to NT password system attacks are

- *Log-on attempts.* In this kind of attack, the hacker attempts to break in by guessing a legitimate user's password by logging in from the user's workstation or from a remote location. In this situation, weak passwords provided the attack opportunity.
- *Captured password attacks.* Typically, a sniffer program captures the hash value of the NT user password from the authentication traffic portion of a user log-on operation.
- The user account database in the registry.

In enterprises in which poor password selection is persistent, NT allows administrators to select passwords for certain users or user groups. The administrator should work with these individuals to select passwords that are easy to remember, expeditious, conform to enterprisewide security policy, and effective. This may be the most feasible way to ensure the use of strong passwords. But if this is contrary to organization policy, run a password-guessing program, such as L0phtcrack, designed to pinpoint weak users in NT environments. As a best practice and to ensure user buy-in, the enterprise should be informed that such password-checking activities will be performed on a regular basis against randomly selected users. When weak-password users are identified, they can be referred to remediation that is geared toward enabling users to consistently create passwords that conform to enterprise password policy.

Log-on-attempt attacks can be virtually eliminated through reasonable password complexity, lifetime, and account locking. Thus, each administrator must develop criteria for password complexity by specifying password length, age—how often to renew—and password uniqueness, or how often new passwords must be created or a favorite one recycled before it can be used again. Generally, eight-character passwords should be used, formulated from randomly generated lowercase alphabetic or alphanumeric characters to create mxyzptlk or er22fo44, respectively, for example. Passwords should also be renewed every 30 days and existing passwords recycled every fifth time. Or, four unique passwords must be developed before you can go back to one that's already been used.

A key feature in NT for thwarting log-on-attempt attacks is password locking, or account lockout. After succeeding in a log-on attack, a hacker typically must guess the password before account lockout is triggered. NT allows users several unsuccessful log-on attempts before being locked out of the system. Once locked out after reaching the lockout threshold, the user must wait 15 or 30 minutes before being allowed another log-on attempt. The count, which allows bad-log-on attempts, is also reset after the lockout duration. Again, the criteria that work best for your enterprise should be determined. However, the following should be used as guidelines in configuring the account-lockout feature:

- Lock out after five unsuccessful log-on attempts
- Reset count after 30 minutes
- Lockout duration of 30 minutes

To enforce your user password policy, NT offers a feature called PASSFILT, a special program that rejects a user's password if it doesn't meet the defined parameters established by the administrator. PASSFILT is not a silver bullet but when used in conjunction with, say, a password-guessing program provides an effective methodology for enforcing the enterprise's password policy.

One of the first accounts that should be disabled is the guest account. Another account to deactivate is the Null session user. The Null session allows anonymous users to log on to list domain user names and to enumerate share names. Eliminate anonymous log-ons through Null sessions, which is the source for network incursions and exploits. Finally, rename the password for the administrator account before you tackle any administrative duties. In keeping with sound password-creation conventions, make sure that the password is at least eight characters in length and derived from randomly generated alphanumeric or alphabetic characters.

Adding Security Measures in Layers

Although NT, in many respects, is different from UNIX/Linux, security can also be instituted in layers in much the same way (Figure 10–6). MD5 and Tripwire

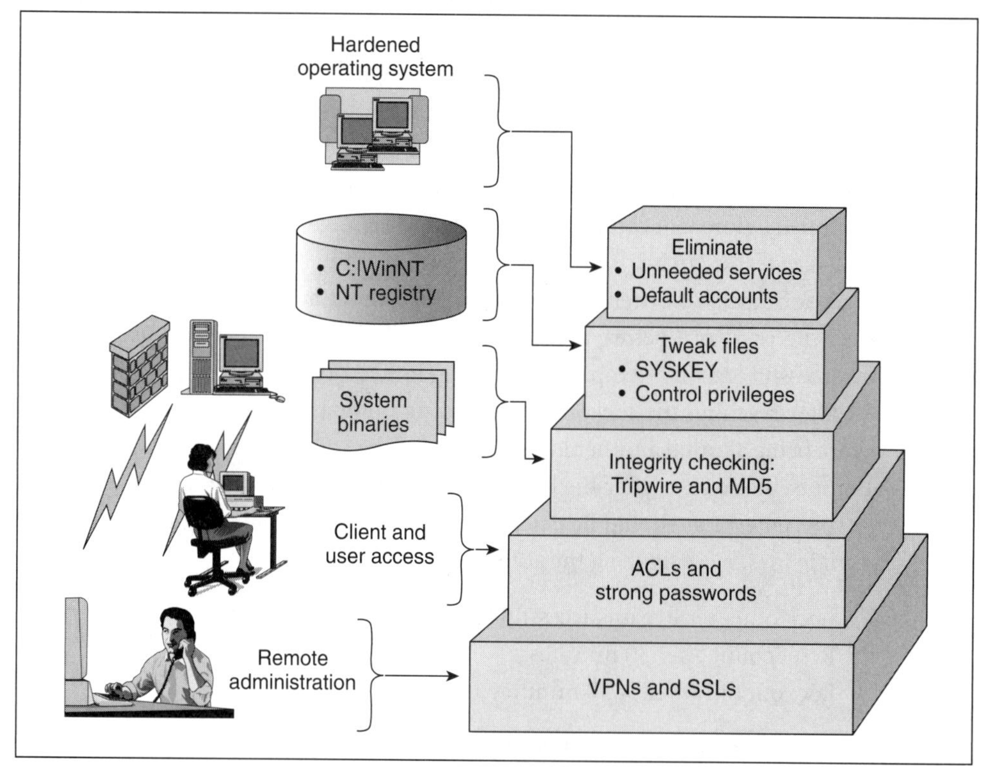

Figure 10–6 Layering security in NT operating environments

can be used to monitor changes in system binaries and related files. SYSKEY is used to protect stored passwords in the (user account database in the NT registry). A VPN or SSL can be used for remote administration or user applications.

Summary

The guidelines in this portion of the chapter are a good point of departure for realizing effective security in operating environments. IT managers can take many other precautions to ensure optimal security in enterprise networks. Although a number of freeware and commercial sources, including books, address the NT security question, one of the most comprehensive freeware documents available for NT security is "Windows NT Security Guidelines," available by downloading from www.trustedsystems.com. The NSA commissioned Trusted System Services to develop this unclassified study for the agency and hence the general public. Another source is "Securing Windows NT," which can be obtained from www.phoneboy.com/fw1. Finally, the NT Server Configuration Checklist from Microsoft is also a good source for NT security.

CHAPTER ELEVEN

Building a Security Architecture

With your infrastructure sufficiently hardened, you now have the foundation for building resilient security architecture. The important point to note about the information in Chapter 10 is that you don't have to be a specialized security professional or hire a security guru to institute the commonsense measures presented. Most important, those measures can be implemented relatively cost-effectively with regular IT/networking staff. The real cost to the enterprise, however, is the cost of *not* instituting the best practices into your regimen of networking activity. Or in practical terms, what is your exposure to risk?

The focus of Chapter 11 is to review the options for building a security architecture and map them to your particular infrastructure and ultimately the enterprise. In other words, we look closely at the architecture as suggested by the IT security functional model (see Figure 10–4). How should the firewall be deployed or tweaked to maximize its effectiveness? Now that my infrastructure is hardened, what steps can I take to track, or monitor, the incidence of vulnerabilities in my network? How do I design my network and/or institute security measures to maximize their efficiency and resiliency in the face of a growing threat? How should these measures be interspersed throughout the enterprise for optimal protection of information assets and computing resources? What is the best way to deploy security for point-to-point communications? And, finally, how do I authenticate the user base and gain user support and buy-in for the authentication system chosen for the enterprise? This chapter addresses those questions, discussing cost-effective solutions that include best practices for deploying comprehensive security architecture.

Firewall Architecture Deployment, Controls, and Administration

For any given firewall, a body of information is available, including published books, white papers, and/or articles that are written for the expressed purpose of optimally deploying your firewall. This information is concerned primarily with properly configuring your firewall's rule base. In the case of stateful inspection firewalls, the rule base, or access rules, must be configured in a certain sequence in order for the firewall to perform correctly. Stateful inspection firewalls enable network access by accumulating key information, such as IP addresses and port numbers, from initial packets into dynamic state tables, to decide whether subsequent packets from a session will be granted access; hence, the name "stateful inspection." Ultimately, the rule base controls access. The stateful inspection activities occur only if the connection is allowed in the first place. This section focuses on deployment strategies, user access, and administrative controls instituted to ensure an effective firewall implementation.

Types of Firewalls

The architecture you deploy for your firewalls depends on a number of factors, such as the level of security required, risk tradeoff, performance requirements, interfaces desired, access controls, network resources, and general application and user requirements. The deployment also depends on what you are willing to spend. In fact, the architecture ultimately deployed must necessarily balance cost against the other factors.

The cardinal rule for implementing any firewall architecture is to never allow *untrusted* external networks *direct connections* into the *trusted* internal network environment, especially if the external network is the Internet. A corrollary is to keep *valid* internal IP addresses from traversing untrusted external networks. Three of the most effective firewall architectures that incorporate and function on these premises are multihomed host, screened host, and screened subnet, or sandbox.

Multihomed Firewall Host

A multihomed host has more than one physical interface or network interface card (NIC) installed. A dual-homed host, with two NIC cards, is the most common example of this type of architecture. In this scenario, one NIC card is connected to the external, or untrusted, network; the other NIC card, to the internal, or trusted, network. A trusted network may be defined as an internal enterprise network or as a network involving a business partner: an extranet. A trusted network, therefore, is best defined as one that shares the same security policy or that implements security controls and procedures that yield an agreed on set of common security services and precautions.

In a dual-homed host, IP packet forwarding is disabled between the two NIC connections by default, so that traffic originating from external, untrusted networks never directly connects into the trusted, or internal, network environment. Typically, dual-homed hosts complement application proxy firewalls, which terminate and reinitiate user connections after fully inspecting all packets that are seeking entry into the trusted networking environment.

Traffic from the external, untrusted network is received by the proxy firewall through the *external* NIC connection. The proxy disassembles all the packets received, filters out suspect or risky commands, recreates the packets, and, on determining that they are valid, forwards them through the *internal* NIC to a destination within the protected network (see Figure 11–1).

In effect, network traffic from an external cloud never directly traverses into the trusted mission network. On the other hand, the proxy never forwards internal IP addresses of outbound traffic, which is received from the internal NIC connection. In this manner, valid internal IP addresses are *translated* by the proxy such that the only IP address ever seen by external, untrustworthy networks is the IP address of the external NIC of the dual-homed host.

When a dual-homed proxy host works in conjunction with the packet-filtering rules of a router, the resulting architecture provides two effective layers of security. The firewall proxy skillfully captures any suspicious packets that somehow slip through the router's packet-filtering rules (see Figure 11–2). In summary, this firewall architecture is ideal for networking environments with high security needs. It ensures not only that untrustworthy network traffic never directly connects into protected networks but also that application-borne attacks, which typically slip through packet-filtering rules, will also never reach the protected networking environment.

Screened Host Architecture

Screened, or bastion, firewall host architecture uses a host, or bastion host, to which all external hosts connect. Instead of allowing outside hosts to connect directly into potentially less secure internal hosts, these connections are routed

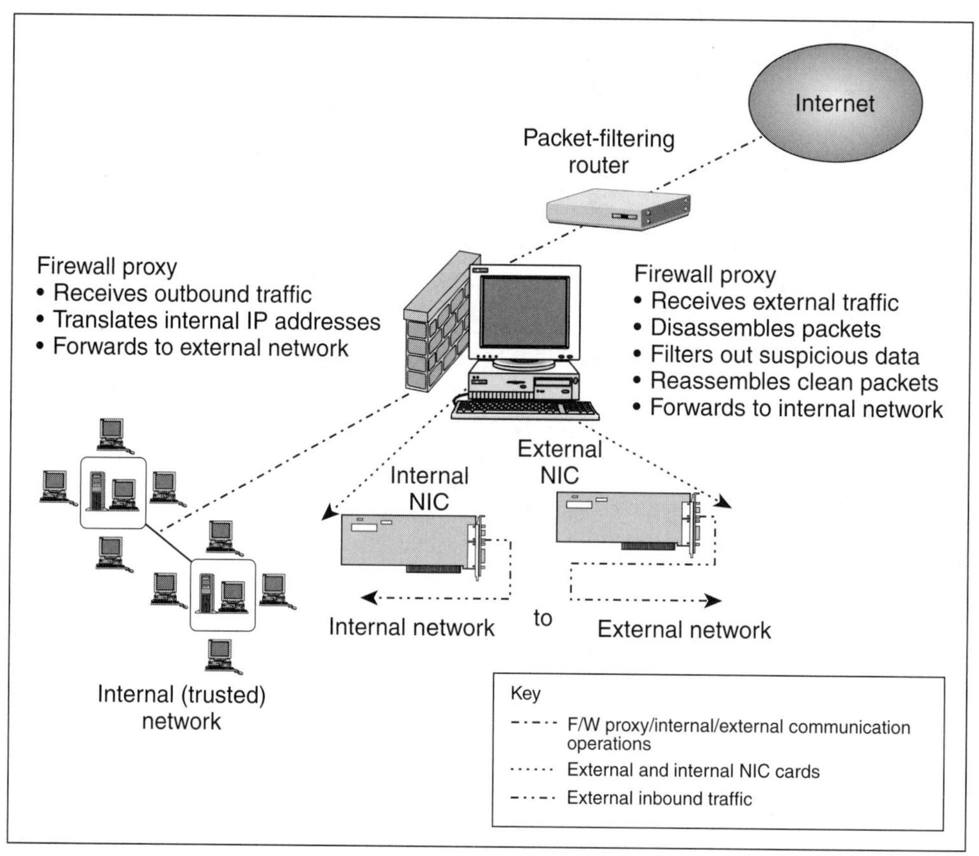

Figure 11–1 Dual-homed firewall proxy host

through a screened firewall host. Achieving this functionality is fairly straightforward. Packet-filtering routers are configured so that all connections destined for the internal network are routed through the bastion host. When they enter the host, packets are either accepted or denied, based on the rule base governing the firewall.

Bastion host architecture is suited for organizations requiring a low to medium level of security. Packet-filtering gateways make decisions on the basis of addressing and port numbers. Application-level attack signatures, usually buried in the payloads of packets, often sneak through the defenses presented by the packet-filtering rules. This is the main argument in favor of proxy-based firewalls, which are effective against application-level attacks.

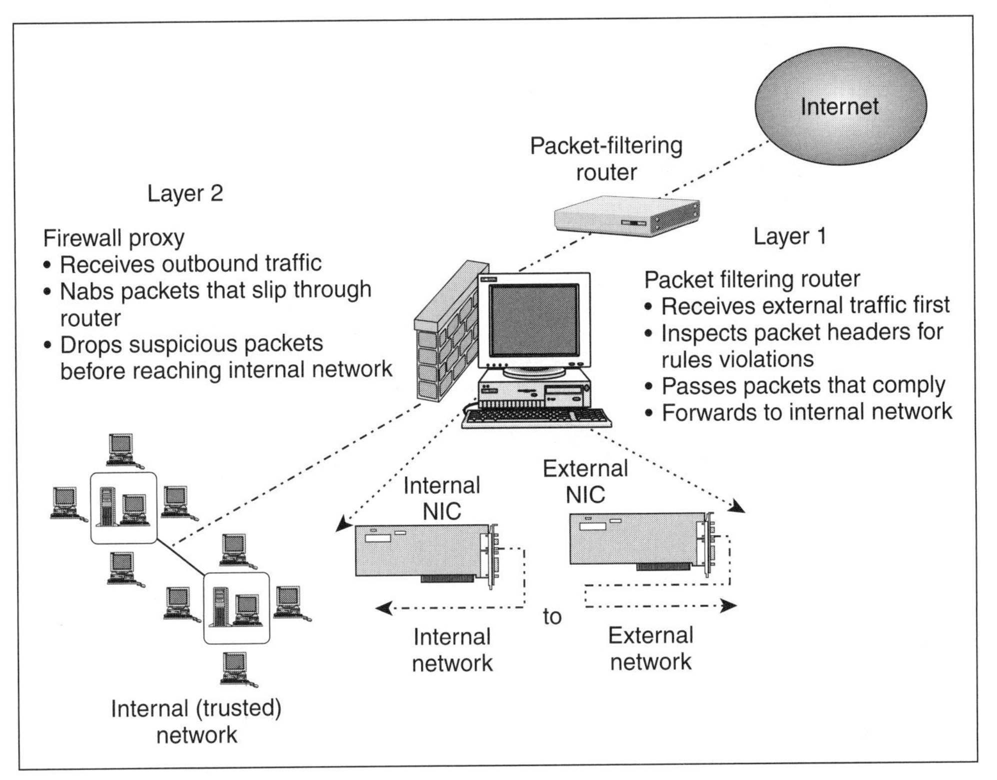

Figure 11–2 Multilayer security example

Generally, packet-filtering firewalls offer faster throughput. Thus, for greater performance, security managers are willing to trade off a lower level of risk for the performance gains. Nonetheless, certain high-performance packet-filtering—stateful inspection—firewalls, such as Cyberguard's firewalls, achieve excellent performance levels with their proxy features in effect.

In summary, the bastion host is an effective alternative when security requirements are low to medium. It also prevents direct connection from external, untrusted sources into protected internal environments. If greater security is required, the dual-homed firewall host may be the better solution, but be prepared to trade off network performance in the process.

Screened Subnet

In today's competitive business environments, the only perceived difference between you and the competition could be the services you offer. This is one of the main drivers for e-business. Enterprises are capitalizing on the awesome potential of the Internet to provide cost-effective services to their clients and prospects. The challenge, of course, is doing it safely. Generally, the more you seek to engage your client or prospect online, the more security is needed. For example, if you are planning to provide the usual information found on a Web site, a Web server with hardened operating system and firewall offers a reasonable level of protection.

On the other hand, if you are planning to conduct business-to-consumer (B2C) or business-to-business (B2B) e-commerce that requires database lookup, record creation, and related database operations, a greater level of security safeguards should be instituted to ensure the safety of your internal network from external, untrusted sources. Additionally, if you intend to offer supplemental services, such as e-mail and file transfer support, providing the full complement of services online securely is best accomplished by a screened-off subnet, or demilitarized zone (DMZ). As the name suggests, a DMZ is a subnet that is screened off from the main, or internal, network. This is done to allow enterprises to offer a variety of online services, at the same time protecting the internal network from untrusted, external access. The screening mechanism is usually a firewall or a firewall and a packet-filtering router.

Owing to the potentially devastating threats lurking in the wild and the fact that the majority of attacks originate from internal sources, operating a DMZ today poses a distinct set of obstacles that must be successfully negotiated if acceptable returns are to be realized. Through best practices and effective architecture, your business objectives can be achieved. The most important thing to bear in mind is that the DMZ will be the most likely area to be attacked and therefore a prime source for compromising the protected internal network. The other, more compelling, statistic, based on industry trends, is that the DMZ is twice as likely to be attacked from the internal network. So the precautions that are taken must factor in the internal threat as well. This is ironic, given that the safety of the internal enterprise network is the highest priority. In other words, the internal network must also be protected against itself.

The applications and the systems in the DMZ should never be allowed access into the internal network. In turn, access into the DMZ from the internal network should be for maintenance and administration only, and such access should be restrictive. This leads to an obvious question. If the DMZ is not allowed access into the internal network, assuming that the central database resides within the internal network, how can database mangement system operations, such as lookup and record creation, be initiated?

The answer lies within the architecture. Theoretically, a DMZ network should be built to accommodate all access requirements for a set of applications and systems. When access requirements differ, a variation of an existing set of applications and related systems or the instituting of a new application and system platform may be needed. In other words, whenever access requirements mandate a certain application and related operating system platform, that application and related systems should be attached to their own DMZ network. Moreover, a given level of user access needs might force various levels of *application-level* access within the resulting DMZ.

Let's explore this notion. To safely accommodate the access needs of the general user and the resulting access needs of the application, the DMZ should be divided into two segments: a *public* DMZ and a *private* DMZ (see Figure 11–3). A public DMZ contains all the applications that are intended for general public access. Applications such as the Web server, the mail server, and the FTP server, for example, are made available for all users with homogeneous access requirements.

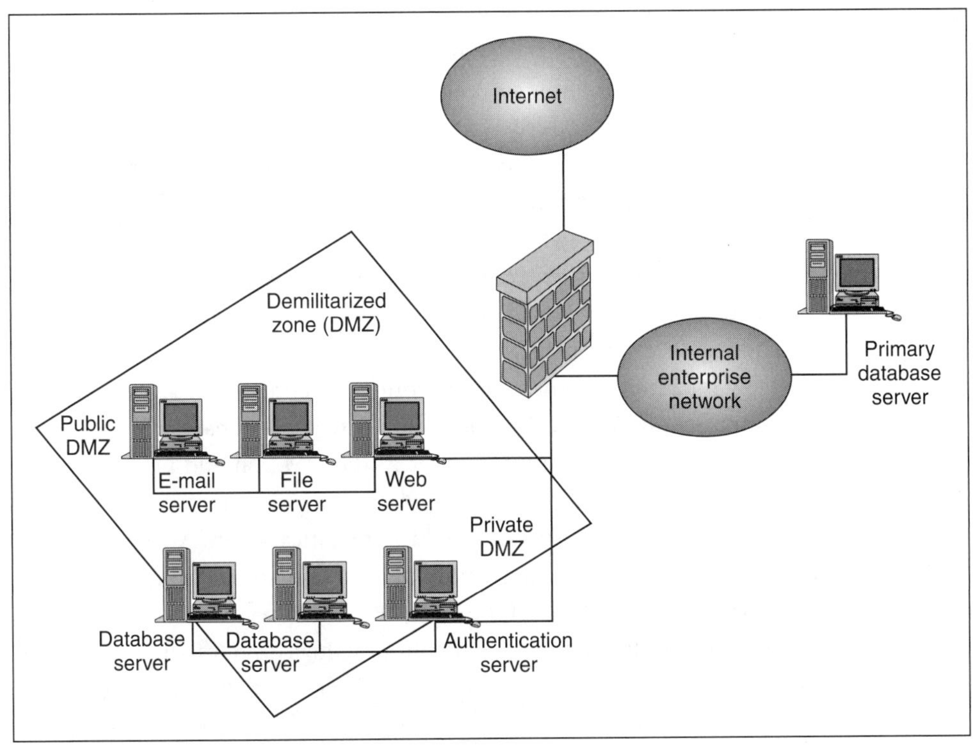

Figure 11–3 Screened-off subnet

In addition, access should be restricted to only those protocols required to access the applications in the public DMZ. In our example, these include HTTP, SMTP, and FTP. All extraneous protocols and default services that are not needed should be disabled or removed to sufficiently harden the application server in question. (See Chapter 10.)

A second, private DMZ is in turn established to accommodate the access requirements of the individual *applications* in the public DMZ. The private DMZ, where the resulting database servers reside, handles database lookup, record creation, and other database requests from the applications in the public DMZ. As a best practice, the private DMZ should never allow access from the general public or from external, untrusted sources. More important, this segregation of duties limits the potential amount of damage and disruption in the event of an attack.

With the appropriate restrictions in place to control external access, restrictions should also be established to regulate internal access. In general, only the individual responsible for maintaining a particular server in the public DMZ—usually a system administrator—is allowed unrestricted access into the server in question. For example, only the system administrator for the SMTP server should be granted access to this server. The Web administrator should be given access to the Web (HTTP) server only, and so on.

Similarly, only the database administrator is allowed to perform maintenance and administration on the database servers on the private DMZ, and the primary database is permitted to perform data transfer operations. Further security gains can be achieved by ensuring that a given system administrator is granted only the specific service and administrative protocols needed to do his or her job. For instance, to maintain the file transfer server, the firewall should enable the system administrator to initiate only FTP connections, along with the requisite administrative protocols to perform related functions.

Overall, each public and private server in the DMZ should implement full auditing and logging capability. As a best practice, a dedicated log server on the internal network side of the firewall should systematically retrieve logs from the DMZ. Ultimately, the dedicated log server prevents unauthorized modifications and ensures the overall integrity of the logs.

On the subject of logging, ensure that firewall logs do not fill up; the typical response is to allow the firewall to simply shut down. In high-available computing environments, this is not feasible. To ensure the 24/7 availability of your firewall in these circumstances, make sure that the one you select allows logs to be written to remote logging servers. For example, Check Point's Firewall-1 allows logging to multiple logging facilities or logging to multiple management servers. In the rare case when management servers aren't available, logging is written to the firewall's local drive. Netscreen Technologies' Netscreen-100, a firewall appliance with no hard disk, allows logs to be written via the UNIX SYSLOG utility to a virtually unlimited number of log servers. Make certain that you are fully aware of how logging is supported by your firewall, especially if it is to protect a highly available computing environment.

As a final recommendation, to facilitate general changes and enhancements to the applications running in the DMZ, this activity should never be conducted in real time directly from the internal network. Instead staging servers within the internal network should be used. Staging servers are configured in exactly the same way as the servers on the DMZ. Any changes to be implemented are first instituted on the staging servers and tested accordingly. Once the changes are verified and the functionality validated, the new modifications may be deployed to the DMZ. If staging servers are not financially feasible, the analysts responsible for the changes must be given access restrictions that are similar to those imposed on the system administrators.

Hardening Firewalls

Out of the box, default settings of firewalls may pose certain vulnerabilities. Some firewalls may be configured to accept routing updates. With routing updates enabled, an attacker can provide bogus routing information during a session and divert traffic to an untrusted domain. Fortunately, this potential vulnerability can be resolved by simply deactivating routing updates, allowing the firewall to use static—regular—routing instead.

TCP source porting and UDP access are examples of other vulnerabilities that may be prevalent in firewall default configurations. For example, these potential vulnerabilities are prevalent in the default configuration of Check Point Software's Firewall-1 version 4.0. With source porting, an attacker uses a trusted port, such as port 80 for Web/HTTP traffic, to mask his or her entry through the firewall. Instead of the typical HTTP packets, the malicious packets used in the attack may be launched against the firewall or hosts that are protected behind its perimeter.

A firewall can counter this attack in two ways. Firewall-1 can be configured to block traffic from suspect sources by initiating connections from well-known sources instead. Also, activate the proxy module for a given protocol/service to nab malicious code buried in the payloads of packets from malicious sources. Like many stateful inspection firewalls, such as Firewall-1, proxy modules are available to work in tandem with stateful inspection modules to ward off payload-borne—application-level—attacks. TCP source porting can mask certain DDoS, virus, and Backdoor attacks. Make certain that the right precautions are initialized with your firewall to block TCP source porting.

Many firewall default configurations deny all traffic unless expressly permitted. In these circumstances, ensure that only the necessary ports and services are allowed through the firewall to support network operations. UDP (User Datagram Protocol), which is notoriously insecure, should continue to be blocked at the firewall. In contrast, the default configuration of Firewall-1 version 4.0 allows DNS updates to transpire through UDP port 53. This could potentially allow attackers to provide bogus DNS entries to perpetrate session hijacking or a diverting of a trusted user host communication to a rogue domain or Web site without the user's knowledge.

Subsequent sessions with the rogue domain will continue indefinitely or until discovered. But in the meantime, much information can be compromised. Firewall-1 can handle this problem easily enough by deactivating DNS forwarding through the firewall configuration's Properties menu.

Finally, make certain that your firewall blocks ActiveX controls and script-based applets. As you know, the potential for malicious scripting attacks poses a huge risk to the internal networks. With these and the forgoing streamlining steps, you are well on your way to deploying an effectively hardened firewall for a critical layer of network protection.

Remote-Access Architecture

This section assumes that you will not use an insecure protocol, such as TELNET, with your remote-administration solution to administer a firewall or a server. TELNET and other insecure protocols should never be used through untrusted network domains for any purpose, let alone remote administration.

The safest way to administer to firewalls and dedicated servers is *locally,* through the attached terminal or console. For local administrative access control, the console workstation can generally be electronically protected through passwords and log-on IDs. In addition, physical precautions can be taken, such as disconnecting and locking away the keyboard and/or locking the door to the server room to limit access. Or a card reader can be installed on the door to control access electronically. Also, idle time-out mechanisms can be used to log out the administrators if keystrokes aren't detected after a certain interval of time. However, if circumstances dictate that administration must take place remotely, physically secure the server in question, and secure remote communications in a particular manner, especially if they must traverse untrusted networks. Under these circumstances, the goal is to avoid communicating in clear text. Depending on the activity, administrators can choose a third-party VPN—depending on the operating system, secure socket layer (SSL) or secure shell (SSH)—or built-in encryption mechanisms, such as those included with commercial firewalls.

Encryption Options for Administrators

Virtually all firewalls have built-in encryption functionality for remote administration. But they differ in the encryption algorithms, key length, and encryption key management systems that are used. Most vendors support the emerging standard Internet Key Exchange (IKE), driven by the Internet Engineering Task Force (IETF). IKE is a set of rules that specify how two end hosts—the systems administrator and the firewall gateway/server—negotiate for exchanging keys that will be used to encrypt sessions. IKE makes sure that the keys match and are authentic before communications get under way. Secure hash algorithm (SHA-1) and message digest 5 (MD5) are the hashing algorithms used to ensure that encryption keys are authentic, or dispensed between the parties that are expecting to be involved in the session.

In addition to encryption, which adds privacy to the communication, the system administrator should use strong authentication, or two-factor authentication, to access the system. Two-factor authentication refers to something the individual must *own* and *remember* to log on to systems. Smart cards or tokens, which are usually not bundled with the firewall, are the most common form of two-factor—strong—authentication. These hand-held devices, the size of credit cards, provide the administrator's (user's) log-on ID. In addition, the administrator must enter a personal identification number (PIN) through a challenge/response mechanism, which in turn will issue a one-time password. A password is created at set intervals from a random-password generator built into the system. After passing the challenge/response, the administrator (user) is accepted as authentic, and access to the firewall server is granted. Typically, smart cards or tokens must be purchased separately from the firewall.

If an enterprise has many system administrators who are on the move, a two-factor authentication system may be the best option for realizing an acceptable level of security. The network's firewall and other servers can be administered from various geographically dispersed locations or from separate offices within a given location. In other circumstances, the log-on ID and password mechanism, coupled with the built-in encryption schemes, would offer an acceptable level of security for remote administration.

Securing Remote-Administration Pipes for Administrators

When you are doing remote administration, services transmit user log-on IDs and passwords in clear text. When crossing untrusted networks, they are subjected to sniffers, connection hijacking, and network-level attacks. Therefore, the protocols and services should never be used to support connectivity when administering to systems remotely. The SSH protocol can substitute for TELNET, rlogin, and FTP in UNIX-based systems. In browser-based systems, SSL is the preferred method. Also, a virtual private network (VPN) can be used for remote administration connectivity, regardless of the operating system used by the server. Each method enables remote log-on attempts and sessions to be encrypted, data to transmit with integrity or without unauthorized modifications, and authentication of system administrators. Additionally, VPNs support the use of smart cards and tokens for a mobile administrative staff.

Administrative functions should also be protected from unscrupulous internal sources. A growing best practice involves using out-of-band management to support administrative functions (see Figure 11–4). With this technique, a separate subnetwork is created to allow access for administrators only. Each server or host, including routers, must be outfitted with an interface, typically an NIC card dedicated for the administrative subnetwork.

If SNMP is popular in your enterprise, creating an out-of-band subnet for the administrators may offer a viable alternative. Working in conjunction with TCP wrappers, available to UNIX and Linux systems, and administrative, file, and directory permissions, available to Windows (NT/2000)-based systems, the out-of-band subnet solution would be a cost-effective means of protecting hosts from internal saboteurs while providing an effective level of security for administrative functions.

Remote-Access Architecture/Solutions for Users

Providing secure network access to both local and remote users is one of the biggest challenges facing IT managers. Moreover, the larger the organization, the more difficult the enforcement of user access security policy becomes, exacerbating

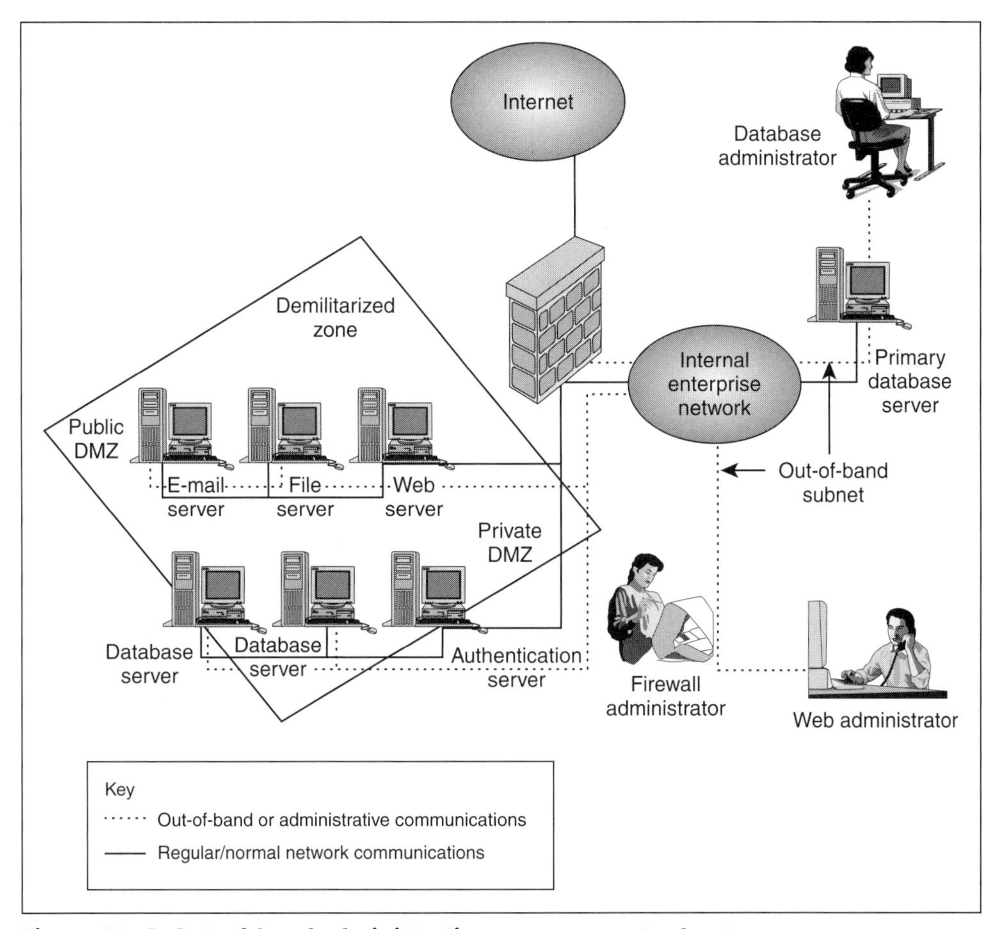

Figure 11–4 Out-of-band administrative management subnet

a potentially tenuous situation. Getting users to consistently use strong passwords with their log-on IDs is at the heart of the problem. The tendency is to reuse easily remembered passwords or to write down more difficult ones. Some users may require multiple log-on IDs and passwords to perform their duties. Under these circumstances, users may find it difficult to adhere to enterprise password policy, especially if it requires them to create a unique password for each log-on ID, say, every 4 weeks.

Furthermore, regardless of whether users use strong or weak passwords, they may occassionally be tricked into providing their passwords to perpetrators pretending to be support staff or help desk personnel. This form of attack, euphemistically called social engineering, can be addressed easily enough through end user security awareness programs and training. The bottom line is that passwords should *never* be provided over the telephone or through e-mail, especially to individuals not known to the user.

Smart Cards and Tokens

Enterprises are turning to alternative, potentially more effective means to achieve an acceptable level of control over user log-on procedures. To ensure that strong passwords are consistently used and changed often, organizations are providing their user base with smart cards and/or tokens with one-time password-generating algorithms. A smart card works with a special reader and can be used for multiple applications, such as PKI (public key infrastructure), building access, or biometrics. In contrast, a token typically generates a unique passcode and displays it in a small window on the device over a set interval, perhaps every 60 seconds. The user combines the passcode with the PIN number and manually keys in the data at the PC keyboard to gain access. The authentication server knows every user passcode/PIN combination to expect in any given 60-second interval. Given a match for that user's passcode/PIN combination, access is granted.

Both smart cards and tokens are designed to work with encryption key algorithms, which ensures that passwords are strong and never transmitted across untrusted networks in clear text. The combination of the built-in encryption and the PIN number ensures that the user in question is who he or she claims to be.

Because of the critical role the PIN number plays in the user log-on process, it should never be written down but instead immediately committed to memory. The big problem with this log-on method is that cards can be lost or stolen. If a card is stolen, which is often difficult to prove, it's likely that the PIN number connected with a user's card may also have been confiscated. In these circumstances, a new card or token must be reissued and the confiscated card deactivated to prevent unauthorized network access.

Deploying smart cards and tokens for user log-on and authentication addresses the use and enforcement issues of strong passwords but may create an equally worrisome situation caused by lost or stolen cards. Smart cards and/or tokens may pose no greater risk to being lost or stolen then regular bank ATM cards. However, with the proper policies and user education programs in place, this problem can be mitigated and/or controlled so that the use of smart cards or tokens could be right for your organization.

SSO Solutions

Two-factor authenticators solve the problem of weak passwords. But how do you cope with the multiple log-on requirements? Enterprises run their businesses on a wide variety of computing platforms, and users must be able to readily access enterprise information safely and securely wherever it resides. Single sign-on (SSO) systems are a viable solution to the scenario of multiple log-ons and passwords.

SSO systems enable users to authenticate once to a centralized log-on (SSO) system and then seamlessly connect into each application platform that has been SSO enabled. Some organizations might have a dozen or more application platforms, with each normally requiring a separate log-on ID and password (see Figure 11–5). Once the user is authenticated to the SSO system, the application platforms that have been retrofitted with SSO capability can be accessed from user desktops.

Vendors of SSO solutions support many, but not all, application platforms. UNIX (HP-UX, Solaris, AIX), NT/2000, NetWare, and certain IBM mainframe

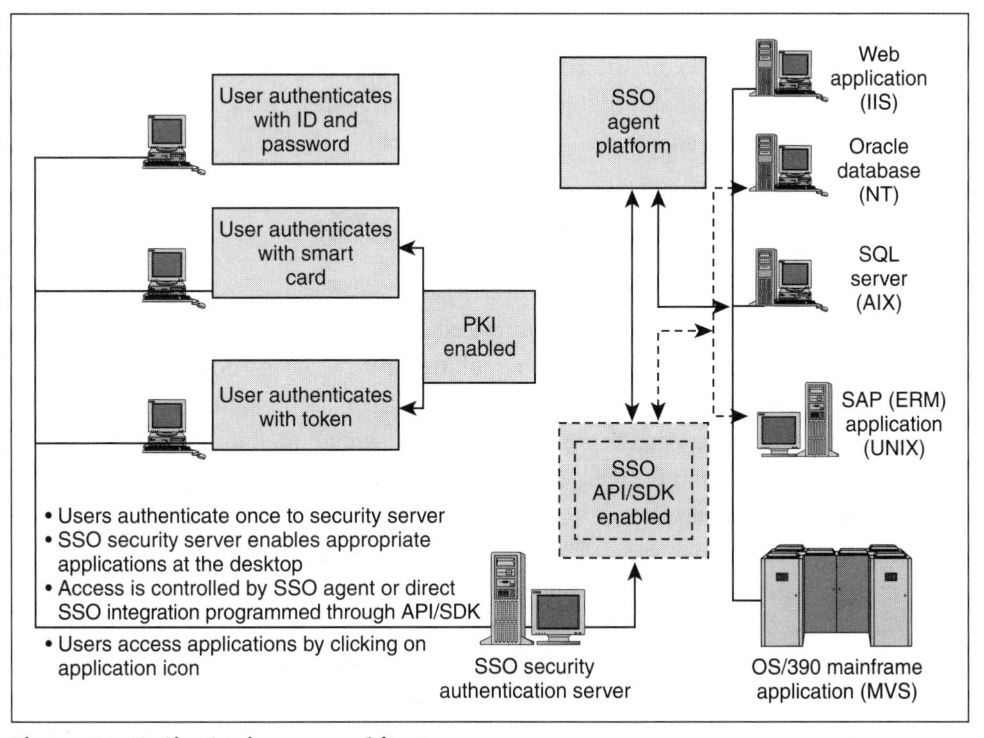

Figure 11–5 Single sign-on architecture

platforms, such as MVS, are supported through ready-made SSO modules or agents that come as part of the standard offering. To develop SSO functionality for other platforms, SSO solutions include a software developers kit (SDK) and/or application programming interfaces (APIs) to create an SSO agent or module. This, of course, requires programmer intervention. RSA's Keon PKI Single Sign On solution, for example, offers an SDK to develop a stand-alone agent for programming a stand-alone SSO module for the application platform in question or a source code SDK to program the SSO functionality directly into the source code of the application. Either way, the application platform will become SSO enabled; however, going the "agent" route may impact performance, owing to the greater level of overhead created for the resulting application platform. In effect, you trade off performance to gain ease of development, functionality, and quicker returns on investment.

In the final analysis, you may determine that the cost of enabling SSO for certain application platforms is too great. In fact, SSO solutions are rarely used to include every application platform supporting business operations, owing to cost of deployment, ROI, and system life-cycle considerations. Nevertheless, SSO solutions offer a viable alternative to the problem of multiple log-on passwords. SSO solutions can also accommodate user authenticator devices, such as smart cards, or PKI for an effective, secure log-on solution.

Vulnerability Assessment Architecture/Solutions

Vulnerabilities are the tell-tale, overlooked, and forgotten windows and doors that hackers violate to gain unauthorized entrance into your enterprise network. The disturbing reality about vulnerabilities is that hackers don't have to be very good to exploit them. With limited skill, a little patience, and perseverance, the average hacker can attack a network with known vulnerabilities and, depending on the exploit, cause major disruption, vandalism, destruction, theft or loss of intellectual property, proprietary secrets, and/or information assets.

One reason an average hacker can be so successful is that powerful tools exist as freeware downloads from popular hacker sites in the wild. (See Chapter 8 for an overview of hacker tools.) Another, more important, reason is that because figuring out what vulnerabilities to exploit depends in large part on a relatively uncomplicated process: determining what operating system, system platform, and services run on a given network. In other words, a hacker who can identify the OS, hardware, services, and specific versions of running services can most likely identify attacks likely to succeed. These attacks come in the form of home-grown scripts, utilities, and related modified programs that are readily available as free downloads from yet other hacker sites. After obtaining OS and related information, the hacker can, with a little research, derive a list of known vulnerabilities.

A program like nmap, the "Swiss army knife of hacker tools," is effective in identifying operating systems, system platforms, and running services through an extensive array of built-in features for probing, port scanning, and OS fingerprinting. To identify the pertinent information, potential hackers may have to probe entire networks or a single host, bypass a firewall, use stealth or undetectable scanning techniques, or scan protocols. Nmap simplifies this process by including virtually all the techniques that would normally require multiple scanning tools that run on multiple platforms. Unfortunately, nmap and similar, but less powerful, tools are in wide use in the wild by hackers and, to a lesser extent, white hats, as well.

For example binfo.c is an efficient little script that retrieves the version of the domain name service running on a remote name server. If the server is running a vulnerable version of BIND, the hacker will also know that the operating system is a flavor of UNIX. With nmap, the hacker could quickly pinpoint the specific operating system and the version of BIND that is running. With a little extra research, the hacker would soon learn that NXT, QINV (inverse query), and IN.NAMED (named) are the three vulnerabilities that are exploited to gain unauthorized control of this particular DNS server.

Attacking the vulnerable version enables the hacker to gain a root compromise wherein system files can be replaced and administrative control obtained. The binfo.c script can be obtained from the Web site http://www.attrition.org/tools/other/binfo.c. Attrition.org is a popular hacker site founded by Brian Martin, who is known in the hacker community as Jericho. Although this exploit can be controlled by applying the appropriate patch levels, the BIND vulnerability is still listed as the number 3 UNIX exploit in the SANS/FBI list of top 20 vulnerabilities (see Appendix A).

As a good practice, IT security managers should obtain a copy of nmap and run it periodically so that they know what a potential hacker is able to see if he or she decides to obtain reconnaissance on your network. If you decide to install nmap, the recommended configuration is a dedicated Linux box.[1]

1. For details on how to download nmap under Linux, go to http://www.insecure.org/nmap/.

The bottom line is that with nmap, hackers can obtain critical information on the OS, system platform, and running services. With such information, a list of known vulnerabilities for the "fingerprinted" operating environment can be obtained without much difficulty. Therefore, the recommended strategy is to eliminate as much vulnerability as possible in mission-dependent networks. A hacker may be able to learn your network's OS, services, and system platform, but if the related vulnerabilities aren't there, you succeed in eliminating the unwanted windows and doors, which are key to a hacker's success.

Vulnerability assessment, or analysis, is accomplished by using specialized tools to determine whether a network or host is vulnerable to known attacks. Vulnerability assessment tools, also known as scanners, automate the detection of security holes in network devices, systems, and services. Hackers patiently probe networks to discover such openings. Similarly, scanners are designed to simulate the behavior patterns and techniques of hackers by systematically launching a salvo of these attack scenarios to explore for *known* vulnerabilities, which could be any of hundreds of documented security holes. The SANS Institute's list documents 600 or more exploitable security weaknesses.

Some scanners, such as Symantec's NetRecon, provide a path analysis, which details the steps an intruder might take to discover and exploit your network's vulnerabilities. When they are discovered, scanners prioritize them in a report. The rating indicates the immediate level of threat potential for a given security hole.

These specialized analysis tools are available from both commercial channels and freeware sources and are used for either network vulnerability assessment or host vulnerability assessment. Nmap, developed by Fyodor; Nessus, by Renaud Deraison and Jordan Hrycaj; and SATAN (Security Administrator Tool for Analyzing Networks), developed by Wietsa Venema and Dan Farmer, are examples of freeware network scanners. SATAN can also be used for host scanning. Nessus tested better than did commercial scanning tools NetRecon and Internet Scanner in the 2001 Network World Vulnerability Scanner Showdown. Nevertheless, NetRecon, ISS's Internet Scanner, and Cisco Systems' NetSonar are all effective network scanners.

For host vulnerability assessment, commercial tools appear to be in more widespread use and acceptance. Leading the commercial offering for host vulnerability analysis are Symantec's ESM, ISS's System Scanner and Database Scanner, and Network Associate's CyberCop Scanner. On the freeware side, you can use nmap, SATAN, and a freeware multipurpose tool called COPS (Computer Oracle and Password Systems), developed by Dan Farmer.[2]

Network-Based Assessment Architecture

Network scanning should be conducted from a single system. Depending on the scanner, the scanning engine can operate under various operating environments. For example, NetRecon version 3 with security update 7 operates under Windows NT 4.0, service pack 3 or greater, or a Windows 2000 workstation or server. The recommended operating environment for nmap is Linux.

Regardless of the scanner's operating environment, the system can scan various operating environments, including UNIX, Linux, Windows 2000, Windows NT, and NetWare. The scanning system analyzes the external side of an enterprise's firewall, servers, workstations, and network devices by launching a variety of attacks and probes to categorize weaknesses (see Figure 11–6).

In the final analysis, the completed scan enables you to determine what havoc an intruder could wreak, what services could crash, or what denial-of-service attacks the network is preconditioned to. Scanners can also help you ensure that the enterprise security policy is resilient in the face of simulated attacks. More important, a scanner can be instrumental in conveying whether your firewall can be penetrated. Of course, a successful penetration may suggest that remediation is in order. As a best practice, network vulnerability assessment should be run at set intervals or when there is a significant change to the network. Depending on the sensitivity of your information assets and the level of open access, you might be better served to conduct network vulnerability assessment on a monthly basis. Otherwise, conducting one on a quarterly basis may prove to be sufficient. (For additional discussion, see Chapter 12.)

Host Vulnerability Assessment

Host-based analysis works from *within* the network, focusing on internal management controls as enforced through the enterprise security policy. Host assessment systems check the system and application settings within the host and compare the data to the rule base in the host assessment system. The rule base is a

---

2. SATAN can be obtained from ftp://ftp.porcupine.org/pub/security. Further information on SATAN can be obtained from http://www.cert.org/advisories/CA-95.06.satan.html. Nessus can be obtained from www.nessus.org. COPS can be obtained from ftp://coast.cs.purdue.edu/pub/tools/UNIX/cops.

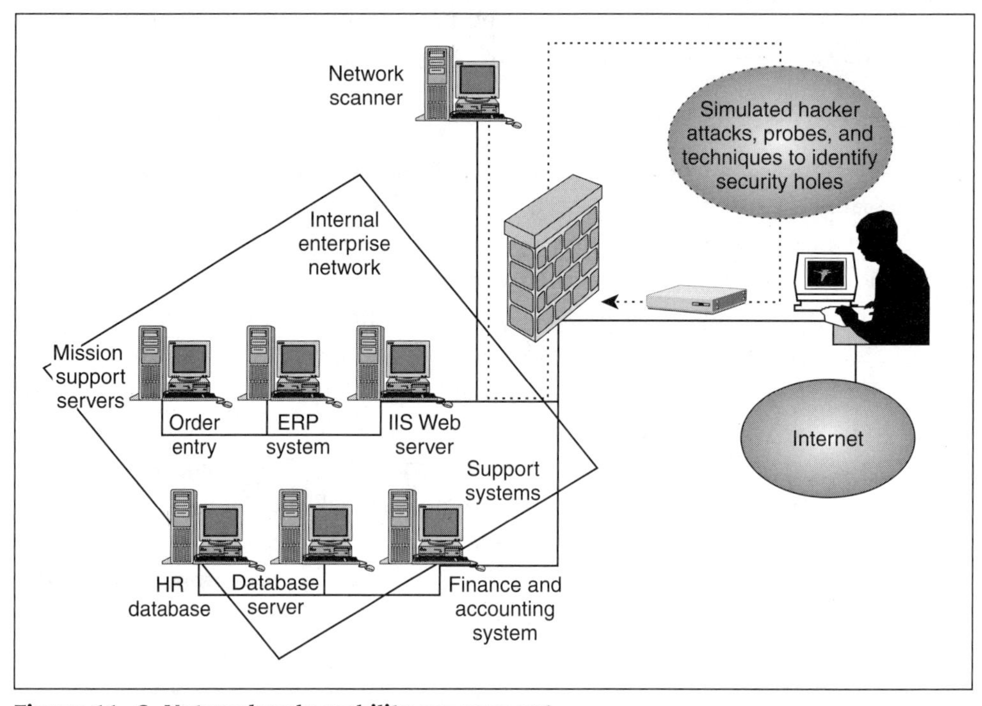

Figure 11–6 Network vulnerability assessment

manifestation of the enterprise's security policy. Therefore, any violation of the rule base is a violation of the enterprise's security policy. The host-based system is designed to run directly on servers, workstations, and applications operating in the network. Because of the sheer number of potential hosts that can exist on a given network, the best strategy in deploying a host assessment system is to start with the critical systems first, such as those featured in Figure 11–6.

In operation, an assessment agent of the host vulnerability system must reside on the host in question. These agents, in turn, report the data to a central host assessment server, which is designated as the central repository, or manager. The administrator interfaces with a GUI to generate reports. The reports tell the administrator which systems are in compliance with the security policy. Systems found not to be in compliance are modified accordingly with the correct settings, security enhancements, and related controls.

Network and host-based assessment are key in building a resilient security architecture. Coupled with infrastructure-hardening procedures, they ultimately provide the foundation on which other security measures, such as intrusion detection systems, firewalls, VPNs, and virus detection systems, should be deployed to ensure proper protection of the enterprise's network. (Additional information on vulnerability assessment is provided in Chapter 12.)

Intrusion Detection Architecture

An intrusion detection system is perhaps the most critical layer in a multilayer security deployment strategy. Intrusion detection systems are designed to render network security in real time and *near* real time. In real-time intrusion detection, an intrusion detection agent continually sifts through *all* network traffic. On the other hand, near-real-time IDS systems take snapshots of traffic on the wire at set intervals. Either way, the IDS sensor must analyze the data stream against a database of *known* attack signatures to determine whether an attack is under way. In effect, the IDS system and its related signature database are only as good as the last known attack. The operative word here is *known*.

A window of opportunity for hackers exists in that interval of time when new attacks originate and the IDS signature database is updated with their signatures. Because of the ongoing threat of new attacks and the immediate time frame before signature databases are updated, some idealists believe that no signature-based IDS system, therefore, is a true real-time security system. At best, signature-database IDS systems can be only *near* real time because *new* attack patterns most likely will not be detected.

Between the point of discovery and when a patch is applied and/or signature database updated, it is up to the IT security manager, not the IDS system, to ensure that the window of opportunity is minimized. For this reason, security administrators must be vigilant in ensuring that IDS signature databases are brought up-to-date as soon as new attack signatures become available. Instead of waiting for your IDS vendor to fingerprint new attacks and to provide them as e-mail attachments or to site downloads, some IDS systems, such as Symantec's NetProwler, allow development of custom attack signatures. For the initiated, this feature will allow network security to stay perhaps one step ahead of—or, depending on your point of view, no more than one step behind—hacker exploits.

Network-Based IDS Architecture

Like an out-of-band administrative subnetwork, a network IDS system should be deployed as a separate network segment. If the IDS system is directly connected into the internal network as regular network hosts and assigned IP addresses, the system would be subjected to attacks. To minimize the potential for attacks, IDS agents or sensors that are responsible for monitoring network traffic should not be configured with network IP addresses. This precaution would render the IDS agent virtually undetectable while enabling it to work in a *stealthlike* mode. But if IP addresses are not provided to IDS agents, how is remote administration handled? More important, can remote administration be supported from a central point?

In all instances, an IDS sensor should be a stand-alone unit configured as a dual-homed host. One network interface card (NIC)—for analyzing all the packets traversing the wire—should be configured in *promiscuous* mode, examining every packet on the local segment while operating undetected in a stealthlike manner. The other NIC card should be configured with an IP address to facilitate IDS reporting and remote administration. With this architecture, the IDS network segment can be controlled from a centralized remote location while operating virtually undetected (see Figure 11–7).

Although the figure shows the DMZ IDS agent connecting directly to the IDS, bear in mind that this is a conceptual diagram. As a best practice, any IDS agent that monitors an external network segment, such as a DMZ, should report back through the firewall. Reporting through the firewall helps restrict access and ultimately increases security to the overall IDS network. Similarly, the IDS agent that is attached to the external Internet connection for monitoring attacks on the external Internet router should also report back through the firewall to the management console.

One important caveat, however: When monitoring external hosts on the Internet side of the firewall, carefully configure IDS agents to avoid unnecessary alarms. This minimizes the alerts and being forced to respond to false positives. Too many false positives run the risk that IDS would eventually not be taken too seriously.

After configuring and deploying the IDS agents appropriately, a separate IDS network segment is created. Ideally, an IDS agent should also be connected to the IDS subsegment to monitor it for attacks, mainly from the inside. However, if proper access policies, OS hardening, and physical security, are established, the risk of attacks from within the network should be minimized.

A separate network-based IDS deployment scheme, with individual analysis and reporting interfaces, offers other benefits as well. Because the IDS analysis and reporting mechanisms are on separate interfaces, the overall performance of the IDS is improved. Further, if a bandwidth-consuming DOS attack hits any of the segments guarded by the IDS sensors, the IDS continues to function.

In general, deploying an IDS system separately prevents the reduction of available bandwidth to enterprise network segments, especially the DMZ, which is usually devoted to public access. If circumstances dictate that the reporting and

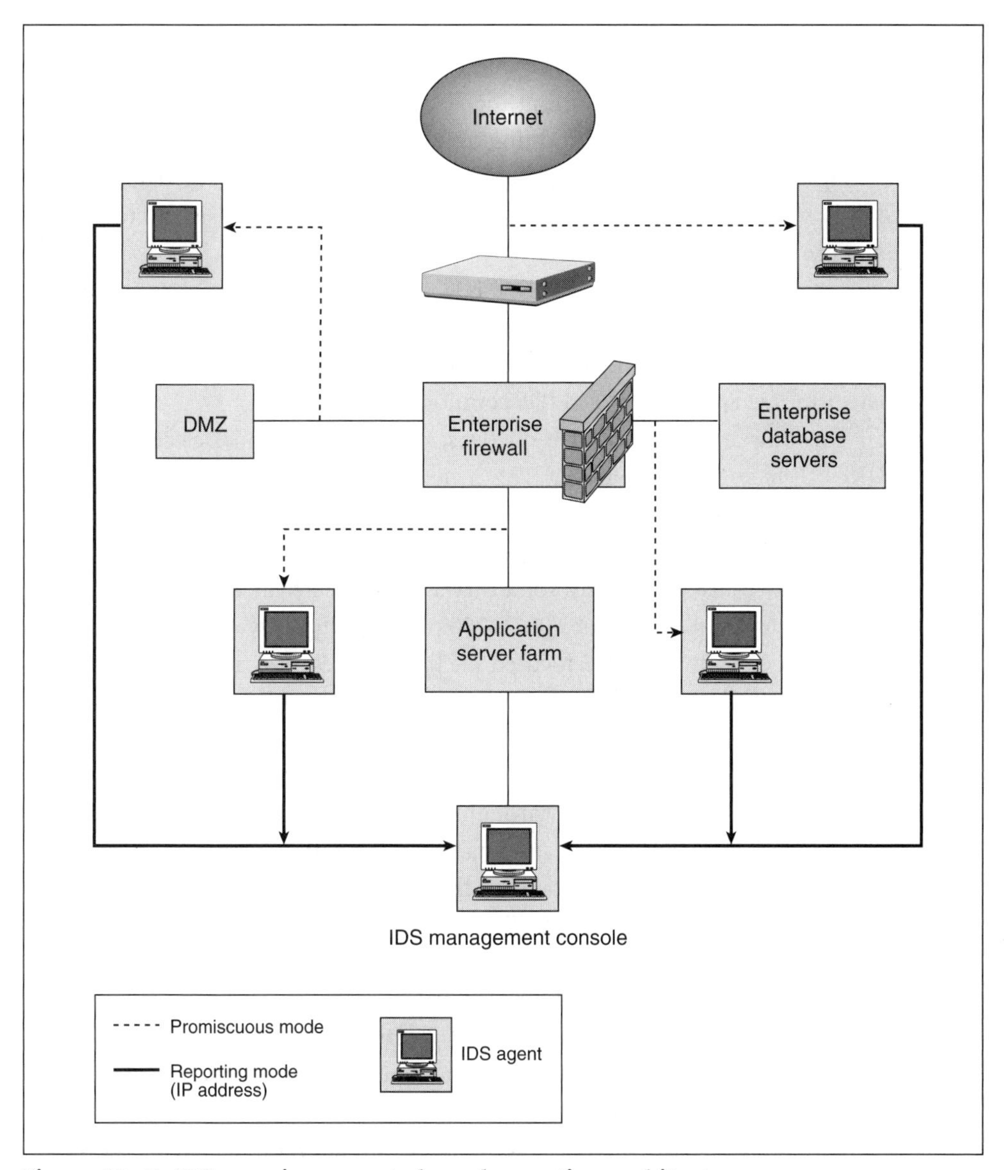

Figure 11–7 IDS promiscuous mode and reporting architecture

control interface must be connected directly to the internal network rather than to a separate IDS segment, security can be maintained by deploying filtering routers and/or switches to restrict access to the IDS management console. The filtering routers and/or switch ensure that only IDS activity, such as reporting, administration, and control, is allowed on the connection.

Finally, communication between the IDS agents and the management console should always be encrypted and authenticated. Additionally, the system clocks of all the hosts supporting the IDS systems should be synchronized to facilitate correlation and the auditing of log data. This includes the clocks of the IDS agents, management consoles, firewalls, and routers. All log data, in turn, should be directed to a centralized log server to facilitate analysis and to prevent unauthorized modifications of logs.

Finally, although network-based IDS systems are effective in seeing all the traffic on a given network segment, they cannot see what is happening on individual hosts. For this reason, a complete IDS implementation incorporates network-based *and* host-based IDS components.

For an excellent perspective on IDSs, refer to "Intrusion Detection Systems (IDSs): Perspective" by Gartner. The review can be obtained from http://www.gartner.com/Display?TechOverview?id=320015. Another excellent review, "NIST Special Publication on Intrusion Detection Systems," is available from the NIST Web site: www.nist.gov. Finally, for a survey of commercially available IDSs, go to http://lib-www.lanl.gov/la-pubs/00416750.pdf; and http://www.securityfocus.com. (SecurityFocus.com is an IDS-focused site that features news, information, discussions, and tools.)

Host-Based IDS Solutions

Host-based IDS systems work through an agent that resides on the host it is monitoring. The agent collects and analyzes information from session events in response to host operations. Because of their proximity to the host, agents are able to analyze activity with great reliability and accuracy but must be configured for full auditing and activity logging. The agent scrutinizes event logs, critical system files, directory permissions, and other audit-capable resources, looking for unauthorized changes and suspicious patterns of activity. Unlike network-based IDSs, host-based IDSs can see the outcome of an attempted attack. When an attack is discovered or anything out of the ordinary is discovered, an alert is issued.

Because the host IDS agent resides on the unit, monitored systems should be configured with additional memory, sufficient disk space, caching, and other resources to ensure acceptable performance. Like network-based IDS systems, communication between the IDS host and the management console should be encrypted and authenticated.

PART IV

Active Defense Mechanisms and Risk Management

In a sense, the subject matter covered in Chapters 12 and 13 is anticlimactic. If you harden your infrastructure, deploy security in multiple layers, increase security awareness in the user base, institute comprehensive management controls, and foster executive management support and buy-in, your enterprise will be well on its way to attaining holistic life-cycle security. The best practices presented in Part IV merely complement, or enable, an optimal e-security posture for the enterprise network. In other words, near-real-time and active defense mechanisms are important, but they rely heavily on the effective dispensation of other e-security components. The interrelationships are not mutually exclusive but instead interdependent. If a vulnerability assessment reveals known vulnerabilities in a critical network component, the fact that every other host is sufficiently hardened doesn't obviate the fact that the network is placed at risk for attack. Vulnerability assessment is effective only if the appropriate steps are taken to eliminate or mitigate the vulnerability discovered.

Chapters 12 and 13 focus on vulnerability and risk assessment management, respectively. Chapter 12 explores how to control those unwanted windows and back doors that crop up during the security life cycle of a network. Chapter 13 covers how vulnerabilities are used to derive the financial threat of potential network attacks.

CHAPTER TWELVE

Vulnerability Management

Vulnerability management can be one of the most challenging e-security areas for IT managers to control, because managing your network's vulnerabilities can be a moving target at best. This chapter reviews the various sources of vulnerabilities. Four main categories of vulnerabilities are discussed to provide insight into the potential scope of the problem. Several recommendations and best practices on how to keep abreast of your network's known vulnerabilities are discussed next. The solutions are practical and key to the overall life-cycle security process. The chapter also further discusses host- and network-based vulnerability assessment. The advantages and disadvantages of each are described to help you better determine how to best deploy complementary security measures that are right for your enterprise's network.

Types of Vulnerabilities

Vulnerabilities in networks and hosts can be created by inherently insecure protocols, from improperly configured network systems and hosts, and from the combination of certain services supporting certain applications. Vulnerabilities also propagate from design flaws in software and related components, from certain networking practices, and from regular applications and ad hoc business applications that may require temporary modifications to computing resources. In general, vulnerabilities can be classified into four categories: vendor-supplied software, system configuration, system administration, and user activity.

- *Vendor-supplied software* may create vulnerabilities through design flaws, bugs, unapplied security patches, and updates.
- *System configuration* vulnerabilities include the presence of default or improperly set configurations, guest user accounts, extraneous services, and improperly set file and directory permissions.
- *Administration-based* vulnerabilities include integration of system services with improperly set options in NT registry keys, for example, unauthorized changes, and unsecure requirements for minimum password length.
- *User activity* can create vulnerabilities in the form of risky shortcuts to perform tasks, such as mapping unauthorized users to network/shared drives; failure to perform housekeeping chores, such as updating virus software; using a modem to dial in past the corporate firewall; and policy violations, such as failing to use strong passwords.

Table 12–1 lists examples of vulnerabilities.

Managing IT Systems Vulnerabilities

Understanding the source of vulnerabilities is an important step in being aware of and ultimately mitigating vulnerabilities in the enterprise network. Controlling them on an ongoing basis requires just as much of a commitment to a methodical review process as the commitment to periodically assess the IT network with host and network scanners.

Many commercial, educational, and public-sector institutions maintain repositories of known vulnerabilities and their related exploits. CERT/CC, Symantec's SWAT, ISS's X-Force, and the MITRE Corporation's Common Vulnerabilities and Exposures (CVE) database are just a few of the most comprehensive databases for tracking vulnerabilities in UNIX, NT/2000, other popular operating systems and applications. (The CVE database, more akin to a dictionary than to a database, is focused on establishing a list of standardized names for known vulnerabilities and related information security exposures.) As a recommended practice, these sites should be monitored systematically to track the incidence of vulnerabilities and their fixes. With up-to-date information, IT managers can institute a proactive system for handling vulnerabilities in their respective enterprise networking environments.

Table 12–1 Major Common Vulnerabilities

Vulnerability	Description
Input validation error	Results when the input to a system is not properly checked, producing a vulnerability that can be exploited by sending a malicious input sequence.
• Buffer overflow (input validation error)	System input is longer than expected, but the system does not check for the condition, allowing it to execute. The input buffer fills up and overflows the allocated memory. An attacker takes advantage of this, skillfully constructing the excess input to execute malicious instructions.
• Boundary condition error	System input exceeds a specified boundary, resulting in exceeding memory, disk space, or network bandwidth. The attacker takes advantage of the overrun by inserting malicious input as the system attempts to compensate for the condition.
Access validation error	The access control mechanism is faulty because of a design flaw.
Exceptional-condition handling error	An exceptional condition has arisen; handling it creates the vulnerability.
Environmental error	The environment into which a system is installed causes it to become vulnerable because of an unanticipated event between, for example, an application and the operating system. Environmental vulnerabilities may exist in a production environment despite a successful test in the test environment.
Configuration error	Occurs when user-controllable settings are improperly set by system/applications developers.

Carnegie-Mellon's CERT/Coordination Center (CC) is fast becoming the de facto standard for incident tracking, alerts, and reporting on vulnerabilities, mainly because of its objective, unbiased handling of security events. Since being commissioned for operation by the Defense Advanced Research Projects Agency (DARPA) in 1988, CERT/CC has handled more than 35,000 computer security incidents and has received more than 2,000 vulnerability reports. CERT/CC publishes security information primarily through advisories, incident and vulnerability notes, and summaries. The bulletins and advisories are disseminated through e-mail received from free subscriptions, Usenet newsgroup bulletin boards, and the CERT/CC site at www.cert.org. For more information on how vulnerability reporting and analysis are handled, go to http://www.cert.org/meet_cert/meetcertcc.html.

SANS Institute is another excellent freeware source for vulnerability alerts; other security services, however, are provided at a reasonable fee. Like CERT/CC alerts, SANS Institute's Resources alerts provide step-by-step instructions on how to handle current network security threats, vulnerabilities, and related problems.

SANS Institute offers "Security Alert Consensus," a weekly compilation of alerts from many accredited sources. Alerts and incident reports are compiled from SANS, CERT/CC, the Global Incident Analysis Center, the National Infrastructure Analysis Center, NTBugtraq, the DoD, and several commercial vendors. Most of these sources have their own individual sites that can be accessed directly. For example, NT security alerts can be accessed from www.NTBugtraq.com. However, through the SANS service, a subscriber can receive summary alerts that are tailored to the needs of his or her particular environment.

Commercial sites worth mentioning are Symantec's SWAT, Internet Security System's X-Force, and Microsoft's security bulletin sites; all offer a variety of information on system vulnerabilities, security articles, and recent developments. The sites also provide a library and allow *free* subscriptions by registering at the site. This results in placement of subscribers on the sites' mailing lists.

As a recommended practice, IT security specialists should register with the Microsoft site to receive security bulletins, especially if their networking resources include Microsoft platforms. Security bulletins are provided for all Microsoft's flagship products, including such Web software as IIS and Internet Explorer. Microsoft security bulletins can be obtained at http://www.microsoft.com/technet/security/bulletin/. Symantec's and ISS's sites can be accessed at www.symantec.com/swat/ and http://xforce.iss.net/alerts/, respectively.

Conducting Vulnerability Analysis

Vulnerability analysis tools generate a snapshot of the security status of a network or a host. In addition to providing an exhaustive search of known vulnerabilities, these tools enable security staff to check for problems stemming from human error or to assess host systems for compliance with the enterprise's security policy. Vulnerability assessments typically proceed in the following way:

1. A particular range of system attributes is sampled.
2. The results of the polling are stored in the tool's data repository.
3. The collected information is organized and compared to the internal library of hacker techniques provided by the vulnerability tool. A host assessment tool, for example, compares gathered information against a specified set of rules that represent the security policy of authorized or allowable user activity.
4. Any matches or variances to the rule sets are identified and listed in a report.

Vulnerability analysis tools are of two major types. One category defines the tool by the *location* from which assessment information is gathered. This scheme refers to either network-based or host-based tools and is the classification that popular commercial and freeware tools are built around. The second, somewhat abstract, category defines the tool by the assumptions regarding the level of trust invested in the tool. Tools in this category are said to be either *credentialed* or *noncredentialed.* These tools incorporate and provide the option to use system credentials, such as passwords or other authentication techniques, to access internal system resources. The discussion in this book applies to network-based or host-based vulnerability systems only.

Network-Based Vulnerability Analysis

Network-based vulnerability scanners reenact network attacks and record the responses to those attacks and/or probe various targets to determine whether weaknesses exist. Network-based assessment tools are an *active defense* mechanism; during the analysis, the system is actively attacking or probing the targeted networking segment.

The techniques these systems use during the assessment are either *testing by exploit* or *inference methods.* Testing by exploit involves launching a series of individual attacks in search of known vulnerabilities. If an attack is successful, the outcome is flagged and the result included in subsequent reporting. Inference methods don't attempt to exploit vulnerabilities. Instead, the system looks for evidence that successful breaches leave behind. Examples of inference methods include checking for ports that are open, system version numbers that illicit queries might check, or requests that seek system status and related information.

Network-based scanners provide two major strengths: (1) centralized access to the security shortcomings and issues of the enterprise network and (2) a network-oriented view of the enterprise's security risks. The first area of strength—centralized access to the security shortcomings and issues of the enterprise network—includes

- Discovery of all operating systems and services that run and/or exist in the networking environment, as well as detailed listings of all system user accounts found through standard network resources. Such network objects are tested for vulnerabilities.
- Detection of unknown or unauthorized devices and systems on a network, which further allows discovery of unknown perimeter points on the network. This is instrumental in determining whether unauthorized remote access servers are being used or connections made to an insecure extranet.
- Ease of implementation and use, because no companion software is required on the host systems that comprise the network.

The second area of strength—a centrist view of the security risks of the enterprise's network—include

- The ability to evaluate network vulnerabilities on the fly by reenacting techniques that intruders use to exploit networks from remote locations.
- Investigation of potential vulnerabilities in network components, such as operating systems, related system services and daemons, and network protocols, especially those that are popular targets, such as DNS or FTP servers.
- The means to assess critical network devices that are incapable of supporting host-scanning software. This refers to such devices as routers, switches, printers, remote access servers, and firewall appliances.
- Postcertification of hosts that have been hardened and/or locked down through proactive security measures. This process involves testing of critical systems, such as file, database, Web, DMZ, application servers, and security hosts, such as firewalls. It also involves testing for configuration errors that would render these servers likely to intruder attacks.

Although network-based scanners are effective security tools, a few caveats are in order. Certain network-based checks, such as those for denial-of-service, can crash the system. Therefore, tests should be conducted in off-peak hours to minimize disruption to the network. These scanners are also platform independent but less accurate and subject to more false alarms. Furthermore, when network-based assessments are conducted, the IDSs can block subsequent assessments.

Host-Based Vulnerability Analysis

Host-based vulnerability systems determine vulnerability by evaluating low-level details of an operating system, file contents, configuration settings, and specific services. Host-based vulnerability systems attempt to approach a system from the perspective of a local user on the system. The objective is to isolate user activities that create security risks in the host. The vulnerabilities typically revealed by host-based assessment involve users' gaining increasing or escalating rights and privileges until a superuser status is achieved. In UNIX systems, this would be a root compromise. In NT or Windows 2000 systems, this would signify illegally attaining system administrator status.

Host-based systems help ensure that a system is properly configured and vulnerabilities patched so that a local user does not gain privileges that he or she is not entitled to own, such as administrator or root privileges. The strength of host-based vulnerability scanners can be divided into three main areas: identification of risky user activity, identification of successful hacker incursions, and recovery detection of security problems that are elusive to network scanners. Identification of risky user behavior includes

- Violations—whether intentional or not—of the organization's security policy.
- Selection of easily guessed passwords or no passwords.
- Unauthorized sharing of a hard disk through default settings—whether intended or not.
- Detection of unauthorized devices, such as modems and related software, such as pcAnywhere. It will also flag unauthorized use of remote access servers that bypass the enterprise firewall.

Identification of successful hacker exploits and recovery includes

- Discovering suspicious file names, unexpected new files and file locations, and programs that mysteriously gained root privileges. Detecting changes in critical system settings in, for example, the registry of Windows systems is another key feature.
- Active hacker activity, such as sniffer programs, seeking passwords and other critical information or services, such as certain scripts, backdoor programs,

and Trojan horses. It also checks for exploits that would take advantage of buffer-overflow conditions in the host.

- For recovery, host-based scanners can create secure MD5 checksums of system binaries to allow security personnel to compare current files to a secure baseline of MD5 checksums created earlier.

Security problems that are elusive or difficult for network-based scanners include

- Performing resource-intensive baseline and file system checks, which are not practical with network-based vulnerability tools. This could potentially require the transfer of the contents of the hard drive of each host on the network to the network vulnerability assessment engine.
- Other scanning that would be difficult for network scanners to perform include password guessing and policy checks, active file–share detection, and search for password hash files.

The main shortcoming of host-based scanners is that their operation depends on a close interconnection with operating systems and the related applications. Because of the potential number of hosts that may warrant an evaluation, host-based security assessment solutions can be costly to build, maintain, and manage. Thus, as a recommended best practice, only mission-critical hosts should be scanned on a regular basis. Other, less critical systems should be sufficiently hardened and kept up-to-date with security patches and revisions. A random sampling of host assessments can then be periodically initiated to ensure that vulnerabilities are controlled and minimized.

In summary, vulnerability assessment systems can reliably spot changes in the security status of network systems, affording security staff an effective remediation tool. Specifically, these systems enable your security staff to recheck any modifications to network systems and vulnerabilities that exist from an oversight in system setup. This ensures that resolving one set of problems does not in turn create another set. Vulnerability assessment systems are instrumental in documenting the state of security at *the start* of a security program or *reestablishing* the security baseline whenever modifications to the enterprise network occur. Finally, when using these systems, organizations should make certain that the systems that require testing are limited to those within their political or management control. Privacy issues must be taken into account, especially when the personal information of employees or customers is included in the information assets that are assessed for vulnerabilities.

CHAPTER THIRTEEN

Risk Management

Risk management is an essential tool for business managers in protecting enterprise assets, especially in such functional areas as finance, manufacturing, and inventory control. Indeed, the concept of risk management is not unique to the IT environment. If the goal is protecting the assets of the organization, if not the organization itself, information and IT platforms have become critical, valuable assets that too must be protected to meet the mission of the organization. In today's Internet economy, many of these IT assets are open to business partners, customers, suppliers and prospects, making the goal of protecting the mission especially challenging.

This chapter explores guidelines for applying risk management concepts and practices to managing *security* risks to the IT environment. If you are the IT system owner responsible for meeting the enterprise's mission and your information system has been opened to valued customers, suppliers, and/or partners, you will find the discussion in this chapter quite useful. The objective of this chapter is to ultimately assist you in making well-informed risk management decisions. The goal is to also provide you with a tool that is traditionally used in the decision-making process by your counterparts in other functional business areas.

The Role of Assessment in Risk Management

What is the risk management process for IT security of network operations and functions? In general, risk management is a process that enables enterprise managers to balance operational and economic costs of protective systems with the desired gain in mission effectiveness. For IT security, risk management involves the ability to balance the cost of protective security measures against the risks to the IT system. IT systems store, process, and transmit mission information. The risk management process must necessarily take this into account and seek a workable balance between the costs of countermeasures and the risks that threaten the information system. No risk management function can eliminate every risk to the information system. But with the appropriate application of risk management practices during the IT system's life cycle, risks with a high likelihood of occurring can be properly prioritized and addressed such that residual risk is acceptable to the overall mission.

A key component in risk management is risk assessment, the most critical phase in identifying and handling risks throughout the life cycle of an information system. The main benefit of risk assessment is identifying and prioritizing risks relative to their potential impact to the mission. Equally important is selecting the appropriate security measure to mitigate the potential effects of risks.

The Process of Risk Management

What is risk? Risk is a function of the probability of a security exploit and the impact that it would have on the organization's IT system and overall mission. Two important factors in understanding risk assessment are the *probability* of the security event and the *impact* of the event on the enterprise's mission. To determine probability, or likelihood, of security exploitation, *potential threats* to the system should be analyzed in connection with the *known vulnerabilities* in the system. To determine the impact of security exploits, *critical elements* of the system must be assessed to appreciate the potential *impact* on the mission. These two factors can also be described as *threat analysis* and *impact analysis*, respectively. Before any threat or impact analysis can be performed, however, the system must be defined, or characterized, to provide the necessary scope of the risk management effort. Several guidelines and related items geared to setting the boundary of the IT system for the risk management activity follow, along with a complete review of threat and impact analysis.

Defining the System Boundaries

Defining the system boundaries establishes the scope of the risk management effort. In addition to providing an understanding of the enterprise's mission and system operations, this step provides information that is essential to defining the risk. This step also provides an understanding of the nature of the mission impact as reflected through the information system and enables the demarcation of the system. For example, information to gather for system characterization includes

- The enterprise's mission
- System processes
- Functional requirements of the system
- System security requirements
- The user community
- The IT security policy
- The system operating environment
- The physical location
- Storage requirements
- Information flows

Once the system has been characterized and boundaries identified, the critical resources and information that constitute the system are determined, completing the playing field of the risk management effort. The information resources or assets can be classified as

- The infrastructure of the information
- Mission hardware
- Specific information assets and related data
- System and application interfaces and resulting connectivity
- System, administrative, and user communities

Threat Analysis

Threat analysis is the process of determining the likelihood, or potential, that a *threat source* will successfully exploit a known vulnerability. To conduct a threat analysis properly, IT security managers should consider four important areas: threat sources, vulnerabilities, existing controls, and probability. Then the likelihood that a threat will exercise a known vulnerability can be determined. Note that without a vulnerability to be exploited, a threat source does not pose a risk. This underscores the importance of controlling the incidence of vulnerabilities in the enterprise information system.

Threat Source Identification

For our purposes, a *threat* is a function of the potential for a threat source to intentionally exploit or accidentally trigger a known vulnerability, resulting in unauthorized access to the enterprise network. A *threat source,* on the other hand, is any circumstance or event with the potential to adversely impact an information system. Threat sources are typically categorized as natural, human, and environmental. All threat sources should be assessed for their potential to cause harm to the system. But obviously, human threat sources in the form of hackers, crackers, and untrustworthy employees command much of the concern and attention of IT security managers. Environmental and natural threat sources are often caused by nature and are accidental or are occasionally caused by intentional circumstances, which are largely unpredictable. But in the final analysis, these types of threat sources can cause as much damage as human sources. However, the probability of your IT system's being harmed from natural and environmental sources is minimal compared to the raw numbers of potential human threat sources in the wild and from within your own networks.

Human Threat Sources

Certain contributing factors to human threat sources are not relevant to the other threat sources. In order for a human to qualify as a valid threat source, motivation, and especially resources, must also be at their disposal. As you know, attacks can be either deliberate or unintentional. Intentional attacks are usually designed to gain illegal network entry for the purpose of compromising a network's integrity, confidentiality, or availability. Unintentional incursions could also result in a compromised network, because once illegal access is gained, the perpetrator may decide to have a little fun. The attack could also be benign, meaning that there is no physical compromise to information, or the attack was launched to circumvent security measures, for example. Table 13–1 provides an overview of the various human threat sources, their motivations for attacking you, and the means by which the attack might be carried out.

How any one of the forgoing individual attackers becomes interested in a potential target depends on many factors. This is where the results of your system characterization activity come into play. Using that information, identify which threat source may apply to each of the system resources, assets, applications, physical locations, connections, subnets, and so on. For example, you may determine that the Web server on your external DMZ caches and temporarily stores credit card information of your clients for a certain period of time. The particular way that your server is configured creates a vulnerability that can compromise this financial data by both *cracker* and possibly *criminal* threat sources. Making this threat source association for each of the integral components of your network,

Table 13–1 Human Threat Sources

Source	Motivation	Attack Method
Hacker, cracker	Ego, challenge, rebellion, or gratification in wreaking maliciousness	Intrusion or unauthorized system access
Criminal	Monetary gain, illegal disclosure of information, theft of financial information, and unauthorized data modification	Intrusion or fraud
Terrorist	Destroy information, exploit it, and/or blackmail	Intrusion and related system incursions
Internal saboteur/intruder	Financial gain, revenge, or capturing intelligence	Abuse, intrusion, and unauthorized access

you can derive a precise mapping of how the individual areas of your network can be compromised by potential threat sources. Once this list of threat agents is compiled relative to the elements produced by your system determination/characterization, you should develop a reasonable estimate of the resources and capabilities needed to succeed in an attack. To give you an idea, this estimate may require the use of a connection into the system, using automated tools to initiate the attack. Or, it could require reliance on insider information of system weaknesses that are not generally known.

One of the critical requirements in threat analysis is to try and maintain an up-to-date list of threat sources. This information is readily available and can be obtained from many government and private-sector organizations. Good government sources include the FBI's National Infrastructure Protection Center (NIPC) and Federal Computer Incident Response Center (FedCIRC). Additional references for potential threat sources are given in Chapter 12.

Vulnerability Analysis

The next step in threat analysis is vulnerability analysis. The objective of this stage is to develop a list of system flaws and weaknesses that could be exploited by potential threat sources. Note that a threat source does not pose a risk without an associated vulnerability to exploit. The purpose of this step is to systematically evaluate the technical and nontechnical weaknesses associated with the network. In addition to using scanners to gather information and assess vulnerabilities, they can be identified through site surveys, staff interviews, and available system and related documentation. The available documentation may vary, depending on the state of the system. For instance, if the system is in design phase, focusing on security policies, procedures, and system requirement definitions can identify vulnerabilities. If the system is in implementation stage, the search should include system design documentation. If the system is operational, your search to pinpoint vulnerabilities should include analyzing the existing security measures to determine whether the controls in use are effective and ultimately able to mitigate risk.

Vulnerability analysis attempts to identify and to assess the level of vulnerabilities on enterprise networks and the potential for their being exploited. A flaw is unlikely to be exploited if there is impotent threat source interest and capability or effective security controls in place. For a detailed review of vulnerability analysis and management, refer to the Chapter 12.

Control Analysis

In this step, the focus turns to assessing the effectiveness of security controls that are implemented in response to the security requirements mandated by the system. The objective is to ascertain whether the security requirements, which are generally identified during system characterization, are being sufficiently addressed. The outcome is to determine what attributes, characteristics, settings, and related features do not satisfy desired security control requirements. Then the necessary remediation can be initiated to bring the system into equilibrium.

Security controls can be classified into three main categories: technical controls, operational controls, and management controls. All are designed to prevent, detect, or recover from a security breach.

- *Technical security controls* are measures that are incorporated into the network architecture. Specifically, they are implemented at the hardware, software, or perhaps firmware levels. Examples are antivirus software, firewalls, strong authentication, Tripwire, and encryption.
- *Operational controls* are those best practices, procedures, personnel, and physical measures instituted to provide an appropriate level of protection for security resources. Examples are security awareness and training, security

reviews and audits, and security plans, such as disaster recovery, contingency, and emergency plans.

Other security controls are designed to support primary security systems. For example, cryptography is used in support of user authentication systems or virtual private networks.

Management controls are simply those that enable enterprise decision makers to manage security and the risks that threaten security. Examples of management controls include the policies, guidelines, procedures, and enforcement provided through the IT security policy, access controls, system security plan, and risk assessment. The most important aspect of management controls is ensuring that the security procedures that are vital to the success and mission of the enterprise are executed in conjunction with management's security directives. In essence, management controls provide the foundation from which the enterprise builds its holistic security apparatus. For this reason, they must necessarily be comprehensive. Therefore, the areas you should explore when devising your controls include enterprise security policies and plans, system operating procedures, system security specifications, and industry standards and best practices.

Probability Determination

The final step in threat analysis involves determining the likelihood that a given threat source exploits a known vulnerability. This is accomplished by deriving an overall likelihood rating. Factors that govern threat probability are threat source motivation and associated capability, the nature of the vulnerability, and the effectiveness of current security controls. The likelihood that a vulnerability will be exploited by a given threat source is described as high, moderate, or low.

- *High:* The threat source is highly motivated and equally capable. Security controls to prevent the vulnerability from being exploited are ineffective.
- *Moderate:* The threat source is motivated and capable from a resources standpoint. However, security measures in place will prevent the exercise of the vulnerability. Or, the threat source lacks the motivation or is only marginally capable of exploiting the vulnerability in question.
- *Low:* The threat source lacks either the motivation or the capability, or controls exist to prevent or significantly curtail the exploitation of the vulnerability.

Completing this step successfully concludes threat analysis, yielding the critical assessment that reveals the likelihood that potential threat sources will breach certain vulnerabilities in your network. Completion of this step concludes a critical component of risk assessment. The remaining part, *impact analysis,* is discussed next.

Impact Analysis

Impact analysis, the next major step in risk assessment, used mainly to determine the resulting impact on the mission in the event that a threat source successfully breaches a known vulnerability. Impacts to the IT system or ultimately to the mission of the enterprise can be qualified with either quantitative or qualitative outcomes. Quantitative outcomes are measured in lost revenue, cost of system repairs, or remediation. Whenever possible, your analysis should attempt to quantify the impacts of potential incursions. Qualitative outcomes express impacts to the mission in terms of either the loss or the degradation of *desired goals* of IT security, such as integrity, availability, confidentiality, and accountability.

Customary practice involves using a rating scale to classify the potential impacts to the IT system. For example, the following rating system could be used to measure system impacts qualitatively.

- *Critical impact:* An attack results in unavailability, modification, disclosure, or destruction of valued data or other system assets or loss of system services, owing to a disastrous impact with national implications and/or deaths.
- *High impact:* The threat results in unavailability, modification, disclosure, or destruction of valued data or other system assets or loss of system services, owing to an impact causing significant degradation of mission or possible staff injuries.
- *Moderate impact:* The breach results in discernible but recoverable unavailability, modification, disclosure, or destruction of data or other system assets or loss of system services, owing to an impact resulting in a transitory adverse impact to the enterprise mission but no injury to persons.
- *Low impact:* The incursion results in unavailability, modification, disclosure, or destruction of data or degradation of system services but does not cause a significant mission impact or injury to persons.

Measuring the impact of a successfully launched attack should be assessed in either quantitative—the preferred manner—or qualitative terms. But what are the advantages and disadvantages of each? One advantage of the qualitative impact analysis is its ability to provide a relative prioritization of the risks and to identify critical areas for remediation of vulnerabilities. One disadvantage is that specific quantifiable measurements are not attainable, thus creating a significant barrier to preparing cost-benefit analysis for desired security controls.

In contrast, quantitative impact analysis facilitates cost-benefit analysis because the magnitude of the impact can be measured—for example, in dollars and cents—and incorporated in the cost-benefit justification for desired security controls. One disadvantage, however, is that if a quantifier other than cost is used, the specific meaning of the quantification may be unclear. Therefore, the resulting interpretation may by default become qualitative in nature. More important, if the quantitative values derived are the product of subjective judgments, which is too often the case, the use of quantitative factors is a mask for qualitative results. The following guidelines may assist you in quantifying the magnitude of an impact:

- An *estimate of the frequency* of a particular threat source exercising vulnerabilities over a certain interval of time, such as one quarter.
- A *reasonable cost estimate* of each potential occurrence of a security event.
- A *weighted factor* based on a subjective analysis of the relative priority of the likelihood of specific security events. For example, given a high likelihood that a threat source will exploit a vulnerability, the weighted factor could be 6. If the event is moderate, the factor might be 4.

Risk Determination

The final determination of risk can be achieved by combining the ratings derived from both threat analysis and impact analysis into one overall risk table (see Table 13–2). For example, if a threat source is highly likely to exploit a given

Table 13–2 Level of Risk Determination

	Likelihood of Threat Occurrence		
Impact	High	Moderate	Low
Critical	Critical	High	Moderate
High	High	Moderate	Low
Moderate	Moderate	Moderate	Low
Low	Low	Low	Low

vulnerability, the source is motivated and capable, and the outcome of such a breach would have a critical impact on the organization, the overall risk to the organization is critical.

Summary

Risk assessment is the cornerstone of risk management. When you are able to map all risk to the IT system and ultimately to the mission to a level of risk determination, the result leads to an understanding of how the potential exploitation of each risk will impact the organization's mission. If the *overall effects* of risks are determined to be high or critical, the necessary steps should be taken to mitigate such risk at the organization's earliest convenience. On the other hand, risks with overall effects rated as moderate to low can be deemed acceptable, requiring little or no response from management.

The discussion in this chapter, one could say, is purely academic. However, if the concepts are applied to real-world circumstances, the critical challenges in applying risk assessment techniques on a regular basis lie in two areas in the overall process. One critical challenge is developing a *realistic* estimate of the resources and capabilities that may be required to carry out an attack. Assessing the capability of a threat source is an extremely important stage in threat analysis. If the necessary due diligence is not devoted to attaining accurate information relative to attack capability, the effectiveness of the assessment is considerably compromised. The other critical challenge involves quantifying the impact of an exploited vulnerability. Again, if the quantification measurement that you use is not realistic, management may have a tough time accepting the recommendations of your risk assessment. Finally, once risks are determined, the appropriate steps should be taken against unacceptable risk.

For an excellent discussion on how to mitigate unacceptable risk and to conduct a cost-benefit analysis to justify acquisition of related security countermeasures, refer to the "Risk Management Guide," NIST Special Publication 800-30, which can be downloaded from the NIST (National Institute of Standards and Technology) Web site at www.NIST.gov.

APPENDIX A

SANS/FBI Top 20 Internet Security Vulnerabilities

Certain host scanners such as Symantec's Enterprise Security Manager (ESM), are able to audit systems to test for vulnerabilities provided by the SANS/FBI Top 20 list. The purpose of this list is to make recommendations for what a host scanner test should look for regarding each vulnerability. The vulnerabilities on this list have been prevalent for some time.

Top Vulnerabilities That Affect All Systems (G)

- *G1: Default installs of operating systems and applications.* OS configurations are audited by flagging unneeded services and services that are not patched, or up-to-date. This provides comprehensive coverage of G1.
- *G2: Accounts with no passwords or weak passwords.* Password strength checking is an important feature. A scanner will alert on many different password settings, such as weak passwords that can be broken through a brute-force attack, weak password length settings, password same as user name, password without nonalphabetic characters, weak password history settings, and so on. It could also identify dormant accounts that should be removed.
- *G3: Nonexistent or incomplete backups.* Especially useful on NT and Windows 2000, a backup integrity feature for Windows NT and Windows 2000 identifies critical servers that are not backing up data on a regular basis.

- *G4: large number of open ports.* Host scanning can typically provide extensive discovery of network services bound to open ports. Security officers can specify forbidden and mandatory services and audit each system against these standards. Network scanners also discover all open network ports.
- *G5: Not filtering packets for correct incoming and outgoing addresses.* Certain firewalls, such as Symantec's Enterprise Firewall, provide ingress and egress filtering by default. G5 is a router and/or firewall activity.
- *G6: Nonexistent or incomplete logging.* An important host-scanning function. A host scanner provides extensive auditing of each system's auditable settings. Security officers can use certain scanners to easily discover audit settings on all systems in their enterprise. Granular audit settings can be discovered.
- *G7: Vulnerable CGI programs.* Usually checks to see whether IIS best-practice policy passes audit of CGI directory permissions, vulnerable CGIs and other script and executable file vulnerabilities on IIS.

Top Vulnerabilities in Windows Systems (W)

- *W1: Unicode Vulnerability (Web Server Folder Traversal).* Tests for a IIS best-practice policy to see whether auditing for this vulnerability flags IIS Web servers that are not patched for the problem.
- *W2: ISAPI extension buffer overflows.* Tests IIS best-practice policy to see whether auditing for this vulnerability flags IIS Web servers that are not patched for this problem.
- *W3: IIS RDS Exploit (Microsoft remote data services).* Tests IIS best-practice policy for vulnerability to see whether IIS Web servers that are flagged for not having patches for this problem.
- *W4: NETBIOS–unprotected Windows networking shares.* Tests for SMB vulnerabilities by listing all shares for each system, shares that are read/writable by everyone, all share permissions, hidden shares, null session connections, and so on.
- *W5: Information leakage via null session connections.* Test for the null session connection vulnerability.
- *W6: Weak hashing in SAM (LM hash).* Test for weak LM hashing in its Password Strength module and LM settings.

Top Vulnerabilities in UNIX Systems (U)

- *U1: Buffer overflows in RPC services.* Test for vulnerable RPC services
- *U2: Sendmail vulnerabilities.* Test all related systems for vulnerable Sendmail services. Some scanners, such as ESM, have default testing facility built in.
- *U3: BIND weaknesses.* Flags all systems that have vulnerable BIND DNS services.
- *U4: R commands.* To identify this vulnerability, test the contents of /etc/hosts.equiv or ^/.rhosts for forbidden parameters that would enable rcp, rlogin, and rsh. Determines when rhost files are modified, created, or deleted.
- *U5: LPD (Remote Print Protocol Daemon).* Test to determine whether LPD settings, such as whether LPD services, are running, LPD user restrictions, LPD dash exploit, and systems running LPD services with no attached printers. Test should also identify LPD services that are not at current patch levels.
- *U6: Sadmind and Mountd.* Test for and identify all running forbidden sadmind and mountd services. In addition, identify the services that are not patched to the current level, using the OS Patches module.
- *U7: Default SNMP strings.* Look for all running SNMP services and audit the services patch level. In addition, detect forbidden parameters and blank parameters within the SNMP configuration files. This vulnerability should also be tested with a network scanner.

APPENDIX B

Sample CERT/Coordination Center Incident Response Form

Incident Reporting Form

CERT/CC has developed the following form in an effort to gather incident information. If you believe you are involved in an incident, we would appreciate your completing the form below. If you do not believe you are involved in an incident, but have a question, send e-mail to: cert@cert.org.

Note that our policy is to keep any information specific to your site confidential unless we receive your permission to release that information.

We would appreciate any feedback or comments you have on this Incident Reporting Form. Please send your comments to: cert@cert.org.

Please submit this form to: cert@cert.org.

If you are unable to send e-mail, fax this form to: 1 (412) 268-6989.

Your contact and organizational information

1. Name:
2. Organization name:
3. Sector type (such as banking, education, information technology, energy or public safety):
4. Email address:
5. Telephone number:
6. Other:

Affected Machine(s) (duplicate for each host)

7. Hostname and IP:
8. Time zone:
9. Purpose or function of the host (please be as specific as possible):

Source(s) of the attack (duplicate for each host)

10. Hostname or IP:
11. Time zone:
12. Been in contact?:
13. Estimated cost of handling incident (if known):
14. Description of the incident (include dates, methods of intrusion, intruder tools involved, software versions and patch levels, intruder tool output, details of vulnerabilities exploited, source of attack, or any other relevant information):

APPENDIX C

Windows 2000 Security/Hardening Plan

The body of knowledge for implementing holistic security measures for Windows 2000 is growing and is readily available. Many organizations either have or intend to migrate to Windows 2000. The good news is that Windows 2000 has implemented potentially excellent features, such as the adaptation of LDAP (Lightweight Directory Access Protocol) for the new Active Directory. Windows 2000 also supports other standards, such as Kerberos for user authentication and IPSec for creating VPN tunnels. The bad news is that Windows 2000 has holes, as do other commercially and publicly available operating systems.

For Windows 2000 security guidance and direction, go right to the source: Microsoft. To its credit, Microsoft has implemented a comprehensive repository of security guidelines and best practices that cover Windows 2000 implementations. Technical documents and white papers are available to provide procedures and/or step-by-step instructions about security measures, best practices, and solutions to attain a safe computing environment.

One such security document is the *Security Operations Guide for Windows 2000 Server*, which provides procedures for locking down the Windows 2000 server to minimize vulnerabilities, as well as best practices for effective management and application of security patches. Guidelines for auditing and intrusion detection are also provided, to round out the in-depth level of information available. The partial guide is directly perusable and accessible from www.microsoft.com/security/default.asp. A complete version of the guide can be downloaded from www.micosoft.com/downloads/release.asp?releaseid=37123.

A wealth of security information on an endless variety of subjects can be obtained from the Windows 2000 Security section, accessible as a menu choice at the top of the Web page of the partial guide. On this part of the site, you can find such documents as

- "Default Access Control Settings in Windows 2000"
- "IP Security for MS Windows 2000 Server"
- "Secure Networking Using Windows 2000 Distributed Security Services"
- "Securing Active Directory"
- "Securing Windows 2000 Network Resources"
- "Step by Step Guide to Configuring Enterprise Security Policies"
- "Step by Step Guide to Internet Protocol Security (IPSec)"
- "Single Sign On Windows 2000 Networks"
- "Windows 2000 Certificate Services"
- "Windows 2000 Kerberos Interoperability"
- "Windows 2000 Security Technical Overview"
- "Windows 2000 Server Baseline Security Checklist"
- Forty additional documents on a variety of subjects, including PKI

To ensure objectivity, you may want to consider third-party commercial and freeware sources for providing Windows 2000 security. Symantec has created a complementary guide to Microsoft's *Security Operations Guide,* demonstrating how to use Symantec's tools to implement the best practices described in the Microsoft guide. The Symantec document can be downloaded free of charge from http://securityresponse.symantec.com/avcenter/security/Content/security.articles/security.fundamentals.html. Also, the NTBugtraq Web site is an excellent independently run source for obtaining related information. Go to http://www.ntbugrtraq.org (Security Focus) to obtain *Securing Windows 2000 Communications with IP Filters.* The site publishes such information on a periodic basis and therefore is a reliable, ongoing source for obtaining critical security information.

APPENDIX D

Denial-of-Service Attacks

Smurf Bandwidth

In a Smurf, or IP directed broadcast, attack, the intruder uses ICMP echo request packets directed to the IP broadcast address of a particular network, which during the attack becomes the intermediary. In addition to the intermediary and the attacker, the other party involved in the attack is the victim. Because the intermediary's network is not a willing participant, the intermediary too should be considered a victim.

In a Smurf attack, the intruder directs an ICMP echo request packet to the IP broadcast address of the intermediary's network. The intruder prefers networks that do not filter out ICMP echo request packets that are directed to the network's IP broadcast address. When the hosts receive the ICMP echo request packet, they all promptly respond with an ICMP echo reply packet at the same time. The source address of the original echo request packet has been spoofed or substituted with the IP address of the intended victim. Consequently, when the hosts respond back with echo reply packets, the response is directed to the host or hosts on the victim's network. The victim is bombarded with a torrent of ICMP echo reply packets. If a large number of hosts are involved, a considerable number of echo request and echo reply packets will be created on the intermediary's and victim's networks, respectively, as the broadcast traffic consumes all available bandwidth. All the DDoS attack tools, except Trinoo, are capable of launching a Smurf bandwidth attack.

SYN Flood

In a SYN flood attack, two events occur: The hacker spoofs, or fakes, the source address and *floods* the receiving applications with a series of SYN packets. In response, the receiving host originates SYN-ACK responses for each SYN packet it receives. While the receiving host waits for the ACK to return, all outstanding SYN-ACKs are queued up on the receiving host's *backlog* queue. SYN-ACKs accumulate until the server's backlog queue fills up or until all available connections are exhausted.

Each SYN-ACK will remain in the queue until the corresponding ACK completes the sequence. However, because the source IP addresses are spoofed, the ACKs never come. The SYN-ACKs would stay on the targeted application indefinitely if not for a built-in timeout mechanism that terminates the connection attempts after a specified interval of time. Usually, these timeout mechanisms are lengthy. Thus, before an attacked or target host can address each SYN-ACK in the backlog queue, a considerable amount of time can transpire, all the while denying SYN requests from legitimate users. SYN-ACK attacks are basic denial-of-service (DoS) attacks.

DoS attacks occur because the attacked host becomes consumed or inundated with processing either *useless session establishment* or *bogus* data. In the case of a SYN-ACK DoS attack, the host tries to respond to bogus address data from illegitimate sources wherein the final ACK never comes. Therefore, requests to establish communication sessions by legitimate users will be ignored by the target host as long as its backlog queue is full.

Stacheldraht

In Stacheldraht, a UNIX-based DDoS attack tool, the master program is contained in mserv.c (master server), and the daemon resides in leaf/td.c. Stacheldraht reportedly runs under Linux but not as cleanly as it does on Solaris UNIX. Stacheldraht corrects one of TFN's glaring weaknesses. When the attacker communicates with the master(s), the interaction is in clear text. In response, this attack tool allows the intruder to encrypt a TELNET-like connection into the master program. In contrast to Trinoo and TFN, which uses UDP, Stacheldraht uses TCP and ICMP to accommodate communications between master and daemon. In fact, the daemon listens for and responds to instructions on TCP port 65000. When communicating with ICMP, instructions are imbedded in

ICMP echo reply packets. The intruder uses encrypted TELNET sessions to communicate to the master program on TCP port 16660.

Encryption is also used between the master and Stacheldraht daemons, sometimes referred to as agents. After successfully connecting to the master, sometimes referred to as a handler, the intruder is prompted for a password. The default password is Sicken, which in turn is encrypted with Blowfish, using the pass phrase authentication. All ensuing communications between the handler and agents are encrypted with this Blowfish pass phrase. A Stacheldraht master or handler can control up to 1,000 agents.

As noted, the master also communicates with the daemons through ICMP echo reply packets. The ID fields in the packets are used for certain values, such as 666, 667, 668, and 669; the data fields, for corresponding plaintext expressions, such as skillz, ficken, and spoofworks. The combination of ID values and data field values is used between master and daemon when the daemon attempts to identify its default handler or master. For example, if a handler's configuration file containing its address is not found by the fledgling daemon after compile time, the daemon reads and contacts a list of default IP addresses of handlers, hard coded in its file, to find one to control it. Therefore, it will send a ICMP echo reply with 666 in the ID field and skillz in the data field to the default IP address of the first handler on the list. On receipt of the packet, the handler responds with an ICMP packet with 667 in the ID field and ficken in the data field. When a connection is made, they will periodically send 666–Skillz and 667–ficken packets back and forth to keep in touch.

Stacheldraht's scariest feature is its ability to regenerate itself automatically. Agent/daemon programs can be directed to upgrade themselves on demand by going to a compromised site to be replaced by a fresh copy of either a Linux or a Solaris version. The command for upgrading itself is .distro user server. Stacheldraht has nearly four times the number of commands—22, to be exact—than Trinoo for manipulating the attack and the attack network. In general, attacks can run for a specified duration from literally hundreds of locations against a target at one IP address or several over a range of IP addresses.

Tribe Flood Network (TFN)

Hackers like this DDoS tool because it can generate a variety of DoS attacks. In addition to a UDP flood, TFN is capable of generating a TCP SYN flood, an ICMP echo request flood, and an ICMP directed broadcast, or Smurf bandwidth attack. In the TFN network, the master functions as the client, and the daemons function as servers. The master communicates with the daemons by using ICMP *echo reply* packets. ICMP echo requests and echo replies are sent and received by the ping command. Instructions to the daemons are embedded in the ID fields and data portion of the ICMP packets. After instructions are received, the daemons generate the specified DoS attack against one or more targets. During the attack, source IP addresses and source ports can be randomized and sizes of attack packets varied. Recent versions of the TFN master may use the encryption algorithm Blowfish to hide the list of IP addresses associated with its daemons.

Tribe Flood Network 2000

TFN2K, a descendant of TFN, operates under UNIX-based systems, incorporates IP address spoofing, and is capable of launching coordinated DoS attacks. It is also able to generate all the attacks of TFN.

Unlike TFN, the TFN2K DDoS tool also works under Windows NT, and traffic is more difficult to discern and filter because *encryption* is used to scramble the communications between master and daemon. More difficult to detect than its predecessor, TFN2K can manufacture "decoy packets" to nontargeted hosts. Unique to TFN2K is its ability to launch a Teardrop attack. Teardrop takes advantage of improperly configured TCP/IP fragmentation reassembly code. In this situation, this function does not properly handle overlapping IP fragments, resulting in system crashes.

The TFN2K client can be used to connect to master servers to initiate and to launch various attacks. Commands to the master are issued within the data fields of ICMP, UDP, and TCP packets. The data fields are encrypted using the CAST encryption algorithm. Another feature of TFN2K is its ability to randomize the TCP/UDP port numbers and source IP addresses.

The TFN2K master parses all UDP, TCP, and ICMP echo reply packets for encrypted commands. Unlike Trinoo, the master server does not require a default password from the intruder at compile time. During the attack, TFN2K enables the intruder to control attack parameters through its *encrypted* command interface. Among other capabilities, TFN2K commands enable attack launches, setting one or more target hosts within a range of IP addresses.

Trinoo

One of the most powerful DDoS attack packages is Trin00, or Trinoo. Trinoo dispenses a UDP flood consisting of large UDP packets that force the attacked hosts to respond with ICMP Port Unreachable messages. ICMP is used to determine whether a machine on the Internet is responding. Ping uses ICMP to check the availability of machines and the validity of the resulting connections.

Preparing for an attack requires coordination between the hacker, who becomes a client to Trinoo's master program(s), which functions as the server, and Trinoo's daemons, which also function as clients. Launching the attack is client/server functionality at its best. The hacker, now an intruder who has succeeded in infiltrating perhaps thousands of Internet hosts, connects with the master programs through TELNET or NetCat. Generally, TELNET creates a virtual terminal that allows computers without the capability to connect to Internet hosts. NetCat is a utility that enables one to read and write data, using arbitrary TCP and UDP ports. From an Internet safe house, the intruder enters a password command via TELNET-TCP connection that starts the attack sequence. The default password, betaalmostdone, traverses TCP port 27665 of the compromised host and initializes the master program. In Trinoo, the master program is contained in a file called master.c, and it expects a password before any communications with the intruder can ensue. The master is instructed to begin preparation for the attack.

The broadcast, or the program that generates the UDP packets that cause the flood, is compiled. The program code is contained in a file called ns.c and is compiled in the host machines that have been compromised by the daemons. Also during compile mode, the daemons hard code the specific IP address of the particular master that controls them. When done, the daemons send a UDP packet containing the string "Hello" back to the master through UDP port 31335. (Note that the intruder—client—communicates with the master on TCP port 27665 and that the daemon communicates with the master on UDP port 31335.) The Hello packet registers the *compiled* broadcast with its respective master.

At this point, all systems are ready. Think of the broadcast as the dam holding the floodwaters back. Trinoo provides six commands for the intruder to control and to manipulate the broadcast: the impending flood. The commands create the size of the UDP flood packets, set the timing of the attack, check for the readiness of the floodgates, target the victim(s), and rescind the waters, or kill the flood. For example, to check the readiness of a broadcast, or daemon-controlled host poised for attack, the intruder instructs the master to send a ping, which is broadcast to all the daemons it controls. The ping is received on UDP port 27444 and is accompanied by the password png 144adsl. The daemons confirm availability by transmitting a pong back to UDP port 31335 of the compromised host containing the master program. Now the intruder can connect to the master(s) and instruct the daemons to send a UDP flood to a specified target or targets. The command the master issues to release the floodgates of the broadcast against a *single* target is the dos command—not to be confused with the IBM DOS command. The syntax of the attack is aaa l44adsl 10.1.1.1. To flood a list of targets, the command used is mdos.

If you understand the underlying services and protocols Trinoo uses, you can gain insight into its characteristics. The intruder communicates with the master via the transport mechanism TCP. This makes sense because the connection has to be ensured and reliable, giving the intruder an advantage while the attack is being planned and carried out. TCP ensures the reliability of data through error recovery and connections that are established by initializing the necessary ports. Besides, the hacker's source IP address and port are spoofed, masking his or her location. In contrast, the fact that the master and the daemon communicate via UDP is also no coincidence. UDP headers are only one third as large as TCP headers, so having lower overhead makes detection of UDP headers more difficult than for TCP headers. Moreover, because UDP is connectionless, meaning that it does nothing to ensure reliable data transfer or a communications link, it's understandable why Trinoo uses ping to verify the availability of daemons. Various TCP/IP services and protocols use ping to verify connectivity.

Furthermore, the fact that Trinoo generates a UDP flood shows how knowledgeable the designers are about TCP/IP and its related protocols. When TCP establishes a connection, it *initializes* the port required to connect to the application in question, at the destination host. In contrast, UDP does *not initialize* any ports at the destination host, because it's connectionless. However, UDP headers do use *destination* port numbers and source port numbers.

Here is where it gets really interesting. ICMP is the *default* service that TCP/IP uses when the destination host must convey information, usually error conditions, to the *source* host. So when UDP packets reach a destination host that wasn't expecting them, it makes sense that the destination host would respond with ICMP Port Unreachable messages. UDP does not initialize any ports, and when a host isn't expecting UDP packets, ICMP intervenes, as it should. In other words, the Trinoo attack relies on inherent functionality, which in this case involves the diligence of ICMP and the connectionless nature of UDP. Recall that the UDP flood is created when the system gets bogged down processing ICMP Port Unreachable error messages in response to an overwhelming volume of UDP packets. So what's the point of all this? Hacker tools are well designed and conceived; consequently, their ability to infiltrate your network should never be taken for granted.

Glossary

BootP (Bootstrap Protocol) An Internet protocol that enables a diskless workstation connected to a LAN to discover its own Internet (IP) address, the IP address of a BootP server on the network, and a file to load into memory to boot the PC. In effect, the workstation can boot without a floppy or hard disk. For obvious reasons, BootP is a favorite service for intruders.

buffer overflow Occurs when a program, especially a utility program, such as a daemon, receives more data input than it is prepared to handle. Programmers build this tolerance into programs to prevent them from crashing. When a buffer overflow occurs, critical areas of memory are overwritten during program execution. Exploiting buffer overflows is a favorite hacker technique; in fact, most incursions are based on this "smashing the stack." Hackers use buffer overflows to slip in and then execute malicious code, which typically results in gaining root access.

directory In computer systems, a certain type of file that is used to organize related other files in a hierarchical tree structure, which metaphorically resembles an inverted tree. The root, or topmost, directory resides at the top of the tree. A directory that is below the root or any other directory in the tree is a subdirectory. The directory directly above any other one is the *parent* directory. File access in a directory is accomplished by specifying all the directories, or directory *path,* above the file. In GUI-based operating systems, such as NT, the term folder is used instead of directory.

egress filtering Used by network managers to prevent their networks from being the source of spoofed IP packets, the potential source of numerous types of attacks. Egress filtering says that on *output* from a given network, do not forward or transmit data that doesn't match a certain criterion. Egress filtering can be implemented on either a firewall and/or routers by configuring these devices to forward only those packets with IP addresses that have been assigned to your network. Egress filtering should especially be implemented at the external connections to your ISP.

finger A service used primarily to target a system for attack. Finger determines whether a particular system has any users; if there are none, the system is "fingered" for attack.

firewall An electronic screening device usually placed between an organization's internal network and an untrusted network environment, such as the Internet. Basic firewall types include packet filters, which operate on the packet level; proxy firewalls, which operate on the application level; and stateful inspection firewalls, which operate and allow connections based on connection parameters in user state tables.

Internet Control Message Protocol (ICMP) Used to handle errors and to exchange control messages. Typically ICMP is used to determine whether a host on the Internet is active and capable of responding. To accomplish this, an ICMP echo request is transmitted to the machine in question. If the host can, it will, on receipt of the request, return an ICMP echo reply packet. Ping is an example of an implementation of this process. ICMP is also used to convey status and error information, including notification of network congestion and other network transport problems.

inetd The Internet daemon, or Internet switchboard operator for UNIX systems. Users access this superserver indirectly. It is one of the most fundamental, powerful background services on UNIX systems. When a UNIX system boots up, inetd initiates requested services by listening in on service-specific ports. It refers to the inetd.conf file to determine which services the server provides on request.

inetd.conf The configuration file for inetd. Inetd.conf contains the specifications for the services that are allowed on a server by system administrators. Inetd.conf is critical to security in UNIX and UNIX-like systems, such as Linux, because numerous security holes are provided by the services and should be removed or locked down.

Ingress filtering A network access-restriction technique that eliminates source address spoofing on the destination side. Source address spoofing allows packets to be forwarded from a network with source IP addresses that aren't assigned to that network's domain. Ingress filters are implemented on external routers of a network domain. On input to an external router on the destination side, the router will not pass the traffic to the network behind it unless it can verify that the packet has originated with the valid IP addresses of the network to which it belongs. In other words, ingress filters prevent the passing of packets with spoofed source addresses. Because ingress filter implementation depends on knowledge of *known* IP addresses, it is practical for smaller ISPs that have knowledge of the IP addresses of downstream networks and for internal enterprise networks.

in.telnetd The service daemon that inetd initializes to support a TELNET session between hosts. When it determines that a particular service is available, inetd instructs the appropriate service daemon—in this case, in.telnetd—to handle the connection. Then in.telnetd starts the TELNET process, including requesting the user's login and password and initializing communication port 23, the port for TELNET.

Java Developed by Sun Microsystems, an object-oriented language that is similar to C++, but simplified to eliminate and to minimize language attributes that cause common programming errors. Java has been optimized to take advantage of the exponential Internet growth. Compiled Java code can run on many platforms because Java interpreters and runtime environments—Java Virtual Machines (VMs)—are available for most of the mainstream computing environments, including UNIX, Windows, and Macintosh.

JavaScript A scripting language for publishing Web applications. JavaScript is a simple programming language that enables Web authors to embed Java-like programming instructions into the HTML text of Web pages. JavaScript however, executes more slowly than Java. Because JavaScript is derived from C, programmers with C or C++ experience will find JavaScript fairly easy to learn. The European Computer Manufacturer's Association (ECMA) recently standardized JavaScript. The interpreter for JavaScript is built into popular Web browsers, such as Internet Explorer.

load balancing Distribution of network traffic to computing resources, such as a firewall or a server, by a device for achieving high availability and resource optimization. The device features a load-balancing algorithm, which dynamically balances traffic in response to decisions derived by monitoring traffic—packets per second—users—client or TCP sessions—and resource feedback, such as firewall CPU utilization, session connections, and so on.

load-balancing algorithm A systemic process of distributing network traffic, based on certain mathematical parameters. Load-balancing algorithms can be based on fewest users, fewest packets per second, fewest bytes per second, percentage of CPU utilization, and so on.

MAC (media access control) address In a LAN environment, assigned to LAN devices to facilitate accurate communications access. For example, vendors of network interface cards burn in the MAC address—usually 6 bytes long and represented as a 12-digit hexadecimal number—into a ROM or EEPROM included in the NIC configuration. MAC addresses are also assigned to other LAN interfaces to establish broadcast, multicast, unicast, and functional addresses, through which related LAN communication is achieved.

MD5 (message digest 5) A cryptographic checksum program used to ensure data integrity in network applications and related systems. MD5 is used to create a fingerprint of software programs, data files, system programs, and so on. MD5 accepts input in the form of, for example, a data, software, or system program message of arbitrary length and creates a 128-bit fingerprint, or message digest, of the input. MD5 ensures data integrity because any given input message string of arbitrary length produces its own unique signature, or fingerprint. MD5 is Internet Protocol Standard RFC 1321.

NAP (network access point) In LANs, the physical location where the network is connected to the Internet. Routers are usually positioned at a NAP, or where data (packets) could flow in two or more directions.

NetCat A general-purpose client utility designed to enable clients to connect to servers. NetCat, available in both UNIX and NT versions, is popular because it allows you to read and write data, using arbitrary TCP and UDP ports. System engineers and programmers find NetCat an invaluable tool for debugging and investigating network services.

null session A utility that allows various services to communicate with one another without the benefit of user passwords and identification. Hackers exploit null sessions to gain unauthorized access to host systems. Through a null session, intruders are able to read password files, user accounts, and network services that are later used to log in as legitimate users.

OpenSSH (Secure Shell) A powerful service that provides a variety of secure tunneling capabilities and authentication methods. OpenSSH, a freeware version of the SSH protocol suite of network connectivity tools, encrypts all traffic, including passwords, to effectively neutralize hacker exploits, such as eavesdropping (sniffing), connection hijacking, and related network attacks. Replace TELNET, rlogin, and FTP, which transmit passwords in the clear and are associated with hacker incursions, such as DDoS and R exploits, with OpenSSH for remote networking connectivity and operations.

operating system kernel The operating system's main module, which loads first and remains in memory while the host is activated. The OS kernel is responsible for the management of memory processes, tasks, and disks. The challenge for manufacturers is to keep the kernel as small and as efficient as possible to ensure a higher level of operating system performance.

Ping A fairly robust utility that can manipulate ICMP echo requests and echo replies in a variety of creative ways. Ping also allows you to specify the length and the source and destination address and to set certain other fields in IP headers.

private demilitarized zone An *external* network used for an internal computing group. For example, an extranet is a private DMZ, a network dedicated for the use of an enterprise's suppliers and strategic partners.

root access In computer systems, the state that provides the greatest level of control over an operating system. Root access allows a user to delete, add, modify, move, and rename files. For this reason, hackers seek to gain root access when infiltrating a system. Root access is typically gained through stages, first by gaining access to a user account(s) and finally by exploiting a vulnerability in the operating system.

rule base A set of procedural statements that translate the enterprise security policy into a base of rules that security measures, such as a firewall or an intrusion detection system, rely on for controlling data communications in and out of networks. In stateful inspection firewalls, for example, rules comprising the rule base must be arranged in a specific order. For rules that are not properly ordered, the desired action(s) may not execute properly when the rule is applied to a given situation.

Sendmail Widely used to implement electronic mail in TCP/IP networks, despite its long history of security problems. Sendmail version 8.7.5 and higher correct the known security problems, and security patches are available to correct the problems in older versions. Current versions also ship with the smrsh (Sendmail restricted shell) program, which is designed to fortify Sendmail against exploits for known vulnerabilities.

shell Technically, the outermost layer of a program. However, shell is more commonly known as a user interface, translating user instructions to appropriate commands understood by the operating system. UNIX systems provide a choice between Motif or OpenLook, shells based on the graphical windowing development system X-Window.

sniffers Programs that hackers use to log information from compromised hosts. Once in place, usually through a vulnerability exploit, sniffers watch all the user activity on the compromised host to trap, for example, user passwords or to capture user sessions. Active packet sniffers, secretly installed on compromised hosts and servers and routers in untrustworthy networks, hijack sessions between client and servers without displaying any perceptible clues. Through active packet sniffers, hackers can intercept, save, and print all transactions transmitted during a given session.

TFTP Used for initializing or booting diskless workstations and copying files. TFTP (Trivial File Transfer Protocol/Service) was implemented on top of UDP to exchange files among networks that implement UDP. Hackers count on administrators to improperly configure this service. When this is the case, TFTP can be used to copy any file on a given system.

TELNET A TCP-based service designed for creating a *virtual* terminal, which allows computers without the capability to connect to the Internet. TELNET servers communicate across port 23; TELNET clients use ports above 1023.

User Datagram protocol A service designed for applications to exchange messages. Typically, UDP is compared to TCP because they both provide data transfer and multiplexing and are inherent functions of TCP/IP. Unlike TCP, however, UDP has no mechanism to ensure the reliable transfer of data and the establishment of connections between hosts. Therefore, UDP is often referred to as a *connectionless* protocol.

Bibliography

Aberdeen Group. *Vulnerability Assessment: Empowering IS to Manage Actual Risk.* Boston: Aberdeen Group, 1997.

Adams, J. "Internet Retailing/Electronic Commerce Update." (Prudential Securities, Inc.) Available online: http://web4.infotrac.galegroup.com. 1999.

Axent Technologies. *Information Begins with Sound Security Policies.* Rockville MD: Axent Technologies, 1999.

———. *Understanding Assessment and Scanning Tools.* Rockville MD: Axent Technologies, 1999.

Bace, Rebecca, Peter Mell, and National Institute of Standards and Technology (NIST). *NIST Special Publication on Intrusion Detection Systems.* Washington, D.C.: Technology Administration, U.S. Department of Commerce, 2001.

Biggs, Maggie. "Good Intrusion Detection Solutions Don't Have to Cost a Bundle," *Federal Computer Week.* Available online: www.fcw.com. 2001.

CERT Coordination Center. "Incident Reporting Guidelines." Available online: www.cert.org/tech_tips/incident_reporting.html. 2001.

——. Intruder Detection Checklist. Available online: www.cert.org/tech_tips/intruder_detection.html. 2001.

———. Protecting Yourself from Password File Attacks. Available online: www.cert.org/tech_tips/password_file_protection.html. 2001.

———. "Steps for Recovering from a UNIX or NT System Compromise." Available online: www.cert.org/tech_tips/root_compromise.html. 2001.

———. "UNIX Configuration Guidelines." Available online: www.cert.org/tech_tips/unix_configuration.html. 2001.

Christiansen, Christian A. *Content Security: Policy Based Information Protection and Data Integrity.* Framingham, MA: IDC, 2000.

Christiansen, Christian A., John Daly, and Roseann Day. *e-Security: The Essential e-Business Enabler.* Framingham, MA: IDC, 1999.

Clark, David Leon. *IT Manager's Guide to Virtual Private Networks.* New York: McGraw-Hill, 1999.

Fennelly, Carole. "Wizard's Guide to Security." Available online: www.sunworld.com/unixinsideronline/swol-05–2000/swol-05-security_p.html. 2001.

Ferguson, P. and Senie, D. RFC 2267 Network Ingress Filtering: Defeating Denial of Service Attacks Employing IP Source Address Spoofing. January 1998.

Frank, Diane. "Training the Security Troops," *Federal Computer Week,* Available online: www.fcw.com/fcw/articles/2000/0410/sec-train-04-10-00.asp. 2000.

Fyodor. "Remote Detection via TCP/IP Stack Fingerprinting." Available online: www.insecure.org/nmap/nmap-fingerprinting-article.html. 1999.

Gagne, Marcel. "Thwarting the System Cracker, Parts 1–6." Available online: www2.linuxjournal.com/articles/sysadmin/003.html. (For Parts 2–6, use 004.html–008.html, respectively.) 2001.

Henry-Stocker, Sandra. "Building Blocks of Security," Available online: www.sunworld.com/unixinsideronline/swol-12-2000/swol-1208-buildingblocks.html. 2001.

———. "Square One: Paring Down Your Network Services." Available online: www.sunworld.com/unixinsideronline/swol-1006-buildingblocks_p.html. 2001.

Koerner, Brendan I. "The Gray Lady Gets Hacked." Available online: www.usnews.com/usnews/issue/980928/28hack.html. 2000.

———. "Who Are Hackers, Anyway?" Available online: www.usnews.com/usnews/issue/990614/14blac.html. 2000.

———. "Can Hackers Be Stopped?" Available online: www.usnews.com/usnews/issue/990614/14hack.htm. 2000.

Korzenioski, Paul. "Scanning for Security Holes," *Federal Computer Week.* Available online: www.fcw.com/fcw/articles/2000/0410/sec-scan-04-10-00.asp. 2000.

Linthicum, David, S. *Enterprise Application Integration.* Boston: Addison-Wesley, 2000.

National Institute of Standards and Technology (NIST). *Risk Management Guide,* Special Publication 800-30, Washington, D.C.: Technology Administration, U.S. Department of Commerce. 2001.

Newman, David. *Super Firewalls!* Manhasset, NY: CMP Media, 1999.

Pasternak, Douglas, and Bruce Auster. "Terrorism at the Touch of a Keyboard." Available online: www.usnews.com/usnews/issue/980713/13cybe.html. 1998.

Pffaffenberger, Bryan. *Building a Strategic Extranet,* Foster City, CA: IDG, 1998.

Power, Richard. "Computer Security Issues and Trends: 2001 CSI/Computer Crime and Security Survey" (VII:1). San Francisco: Computer Security Institute, 2001.

Radcliff, Deborah. "Security, the Way It Should Be," *Computerworld,* July 2000.

———. "Diary of a Hack Attack." Available online: www.nwfusion.com/news/2000/0110hack.html?nf. 2000.

Rekhter, Y, RFC 1918 Address Allocation for Private Internets, February 1996.

SANS Institute Resources, "Help Defeat Denial of Service Attacks: Step-by-Step." Available online: www.sans.org/dosstep/index.htm. 2000.

Schwartz, John. "New Virus Hits World Computer Networks," *Washington Post.* Available online: www.washingtonpost.com/wp-dyn/articles/A37433-2000May19.html. 2000.

Sokol, Marc S. *Security Architecture and Incident Management for E-business.* Atlanta: Internet Security Systems, 2000.

Spitzner, Lance. "Armoring Linux." Available online: www.enteract.com/~lspitz/papers.html. 2000.

———. "Know Your Enemy, Parts I, II, III." Available online: www.linuxnewbie.org/nhf/intel/security/enemy.html. (For Parts II and III, replace "enemy" with "enemy2" and "enemy3," respectively.) 2001.

Sutton, Steve. "Windows NT Security Guidelines." Trusted System Services. Available online: www.trustedsystems.com. 1998.

Symantec. "Enterprise-Grade Anti-Virus Automation in the 21st Century." White paper. Cupertino, CA: Symantec Corporation. Available online: www.symantec.com. 2000.

———. "Responding to the Nimda Worm: Recommendations for Addressing Blended Threats." White paper. Cupertino, CA: Symantec Corporation. Available online: www.symantec.com. 2000.

Thurman, Mathias. "Server Lockdown Locks Out End Users." *Computerworld,* April 2001.

UNIX Insider, available online at www.ITWorld.com/comp/2378/Unixinsider/.

USA Today.com. "Get Ready for 'Code Red' virus version 2.0." Available online: www.usatoday.com/hlead.htm. 2001.

Xtream.Online. "Internet Security." Available online: http://xtream.online.fr/project/security.html.

Index

Note: Page numbers followed by the letters *f* and *t* indicate figures and tables, respectively.

B

C